Exploring the Universe

Science Activities for Kids

Anthony D. Fredericks

Illustrated by Shawn Shea

fulcrum resources

Golden, Colorado

For Bill Kreiger—
passionate educator, dedicated scientist,
and inspirational colleague.

Library of Congress Cataloging-in-Publication Data

Fredericks, Anthony D.
    Exploring the universe : Science activities for kids / Anthony D. Fredericks ; illustrated by Shawn Shea.
        p. cm.
    Includes bibliographical references and index.
    Summary: Provides information about all aspects of the universe: galaxies, the solar system, stars, asteroids and meteoroids, space exploration and more. Includes related hands-on activities and projects.
    ISBN 1-55591-976-6 (alk. paper)
    1. Astronomy—Juvenile literature. [1. Astronomy.] I. Shea, Shawn, ill. II. Title.
    QB46.F765 2000
    520—dc21   00-020371

Printed in the United States of America

0 9 8 7 6 5 4 3 2 1

Book design by Alyssa Pumphrey

Fulcrum Publishing
16100 Table Mountain Parkway, Suite 300
Golden, Colorado 80403
(800) 992-2908 • (303) 277-1623
www.fulcrum-resources.com

Contents

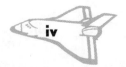

Introduction

A small, shoebox-sized vehicle slowly crawls across the rust-colored surface of a distant planet, testing the soil and analyzing rocks billions of years old. Its driver is 190 million miles away, providing computer commands that guide this self-propelled "car" across an alien landscape.

A group of men and women are sealed inside a compartment smaller than your bedroom as they float hundreds of miles above Earth's surface. They gather important scientific data and conduct valuable experiments in medicine, physiology, and physics.

A nearly 5-billion-year-old star with an internal temperature of 27 million degrees Fahrenheit (F) blazes high in the heavens, providing heat, light, and sustenance to a planet 93 million miles away.

Is this science fiction? Of course not! All this is happening right now. Space probes are making miraculous and incredible discoveries on far-distant worlds, astronauts from many countries are working together in deep space to learn more about our world as well as worlds far away, and millions of celestial bodies are providing scientists and everyday people with magnificent displays, life-sustaining energy, and vast amounts of information about space.

The universe is filled with magical and marvelous discoveries, some being made right now and some just waiting to be made by present-day scientists as well as future scientists (perhaps even you!). Is this an area of science full of excitement, full of amazing information, and full of discoveries just waiting to be made? You bet! Just think, in the time it has taken you to read this section of the book, all the following may have occurred:

- Light from our sun traveled more than 11,000,000 miles.
- A space shuttle in orbit around Earth traveled 295 miles.
- A meteor shooting across the night sky traveled 1,500 miles.
- The planet Jupiter spun 150 miles on its axis.

Many scientists believe that we have only begun to scratch the surface of scientific investigations about space and the universe. Some scientists believe that the next few decades will yield more information about the universe than all the knowledge we have acquired since humans first began looking at the stars.

This book is designed to provide you with exciting and fascinating information about the universe. It offers a number of hands-on activities and projects to help you learn about various aspects of space. You'll learn about the exciting exhibits in your local planetarium; how to locate and observe a wide variety of celestial bodies; and how rockets are shot into space, how they travel, and what they're used for. And you'll conduct experiments similar to the ones astronomers are doing all the time. I invite you to tackle as many of these projects as you wish; of course, the more you attempt, the more opportunities you will have to learn about the universe and all its mysteries. All the activities in this book have been designed to provide you with valuable learning experiences in a variety of areas.

I hope you enjoy these activities and that you will use them as stepping-stones to learning more about the universe in which you live. Space science and space exploration are exciting topics filled with wonder and incredible discoveries. I sincerely hope you enjoy this voyage to the stars—and beyond!

—*Tony Fredericks*

How to Use This Book

T his is a book full of discoveries and full of wonder, a book in which you'll find lots to do and lots to explore. Here you'll get firsthand experiences in learning about and appreciating various elements of the universe. There are projects you can do indoors, discoveries you can make in your own backyard, and "stuff" to learn everywhere you go. This book will not only tell you about parts of the universe, it will show you what our universe is all about. You won't need a lot of equipment or expensive supplies; there are loads of activities, projects, and discoveries for everyone within the pages of this book. The activities are designed to help you appreciate selected portions of the universe and participate in real hands-on learning experiences.

Throughout this book you'll see several symbols such as those below. These symbols identify an activity, experiment, or project for you to try.

 This symbol identifies an activity or experiment.

 This symbol identifies a "Look-for-It" project—something you can do in your own community, in your home, or outdoors. These projects will help you understand and appreciate the universe and how it works.

It is not necessary to complete every investigation in this book. You should feel free to select those activities with which you are most comfortable or in which you are most interested. I sincerely hope you enjoy your journey through the universe, its mysteries, and incredible discoveries.

1 Welcome to the Universe

Write down your age: _____ years old. Add one year to that number and write the new number here: _____. The second number may not seem very important, but think about this: You didn't exist that many years ago. You had not been born or even conceived yet. That number may seem small, but just think about all the things you have been able to do in your lifetime. Think about all the things you have learned, all the things you have seen, and all the places you have been. You've certainly done a lot in the years you've been alive.

Now, think about this number: 4,500,000,000. That's a very, very large number! That's how many years the planet Earth has been "alive" (in existence). During that time it has undergone some remarkable changes. Rocks have formed, primeval seas have ebbed and flowed across vast continents, and dramatic weather conditions have contributed to the geography and structure of our planet. Lots of things have happened to our planet and lots of things have taken place on the surface of our planet in all those years.

Here's one more number to think about: 15,000,000,000. That's 15 billion! That is how many years the universe has been in existence, a number almost too large to comprehend. Astronomers (scientists who study the universe) have calculated that the universe began about 15 billion years ago. One of the most popular theories about how the universe began is known as the "Big Bang" theory. Scientists believe that 15 billion years ago everything was so close together that the universe was a tiny point (known as a singularity) much smaller than an atom. It is also thought that this singularity was billions and billions of degrees hot.

Suddenly there was a spectacular and incredible explosion (the Big Bang) and the universe burst into being. Hot material blasted out in all directions. The heat from this explosion was far greater than that of a nuclear explosion or even the temperature at the center of our sun. Within seconds of this enormous explosion

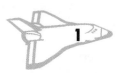

the universe had expanded in size to several million miles. Many scientists (but not all) believe that the universe is still expanding today and will continue to expand throughout its existence. As evidence for this theory they point to the fact that galaxies within the universe are moving away from one another, as if from the force of an explosion. This idea is confirmed by information gathered by radio telescopes throughout the world.

Exploding Universe

You'll need:
9-inch or 12-inch round balloon
black permanent ink marker

What to do:
1. Mark a small black dot on the surface of the balloon.
2. Blow up the balloon until it is about the size of a grapefruit. Hold on to the neck of the balloon so no air escapes.
3. Notice what has happened to the black dot.
4. Now, blow up the balloon some more until it is about the size of a basketball. Again, hold on to the neck so no air escapes.
5. Notice what has happened to the black dot.
6. Blow up the balloon some more, this time until it is about the size of a very large beach ball. Be careful not to blow up the balloon to the point where it breaks.
7. Notice what has happened to the black dot.

What happens:
You will notice that when you first mark the black dot on the surface of the balloon it looks like a single dot. (This is similar to the singularity that existed just before the universe was "born.") When you first blow up the balloon, it expands, just as the universe expanded immediately after the Big Bang. When the universe expanded everything sped out into space. If you look at the black dot after step 2, you'll notice that it has become a series of tiny dots that have moved apart from one another. When you blow up the balloon some more, the dots once again move away from one another. This is what happened to the material that rushed away from the Big Bang when the universe was created. As you continue blowing up the balloon the dots continue to move away from one another. Imagine what you would see if the balloon would never burst and you could keep blowing it up forever. The dots would continue to move away from one another and get farther and farther apart.

The Big Bang created a young universe that was incredibly hot. But as the universe expanded, it began to cool very rapidly. You can experience this type of change also. Stand next to a fire where it is very warm. Move away from the fire gradually. As you move away from the fire, you begin to feel a little cooler. The farther you go from that fire, the cooler you will feel. That is exactly what scientists believe happened in the early stages of our universe.

Fantastic Fact

The temperature of the Big Bang was estimated to be about 1 billion billion billion degrees Celsius (C). One one-hundredth of a second later (much less than the time it takes to snap your fingers), the temperature had cooled to only 1 billion degrees C.

After the Big Bang, as materials moved away from the explosion, gases such as helium and hydrogen were created. These are the two most common elements in the universe to-day. After several million years, these gases collected together into large bodies called galaxies. These first galaxies—known as protogalaxies—were enormous, swirling gas clouds that filled and expanded throughout the entire universe. After about another 1 billion (1,000,000,000) years, true galaxies with stars and spiral arms began to form from these gas clouds (keep in mind that this formation was occurring while the universe was expanding). A galaxy is defined as an enormous group of stars, gases, and dust, all held together by the force of gravity. Astronomers have estimated that the universe contains about 100 billion separate galaxies, each with approximately 100 billion stars. That means that there are about 10,000 million million million stars in the universe.

Far, Far Away

As you might imagine, objects in space are far away from us and from one another. In fact, the distances in space are enormous. It's often hard for us to imagine these great distances, simply because they are unlike anything with which we are familiar. We're more familiar with measurements such as an inch, a foot, a yard, and a mile. But even those measurements are sometimes difficult to imagine. Try the following activity and you'll see what I mean.

Inch by Inch

You'll need:

two blank 8¹/₂-by-11-inch sheets of paper
pencil
index card
dollar bill
copy of *Exploring the Universe*
ruler

What to do:

1. On one blank sheet of paper, draw a line that you think measures exactly one inch. Draw a second line that you think measures exactly one foot.
2. Look at the index card, the dollar bill, and this book. How wide is each of these objects? How long is each of these objects?
3. Write down all your estimates on the other sheet of paper.
4. Using the ruler, measure your two lines and the dimensions of the index card, the dollar bill, and this book.
5. How accurate were your lines and estimates?

What happens:

Even though you are quite familiar with the unit of measurement we call an "inch," you probably encountered some difficulties with your estimates and predictions. That doesn't mean that you don't know what an inch is, it just means that your

perceptions may not match actual measurements, even of objects with which you are very familiar.

Now imagine that you had to figure out how far apart two cities were and you didn't have a ruler or other type of measuring device. All you could do would be to estimate the distance. How much more difficult would that be than estimating the length of a pencil line on a piece of paper? I think you'll agree that it would be considerably more difficult.

Or think of how hard it might be to measure the distance between two objects in space that are very far apart. (Obviously we can't use a ruler or yardstick in space.) Space distances are enormous, and the measurements with which we are most familiar (inch, foot, yard, mile) don't work, simply because the numbers that would result using such measurements would be much too large to comprehend.

Light-Years

Astronomers have developed a unique way of measuring the unbelievable distances in space. They use a unit of measurement known as the light-year.

Although the term *light-year* sounds as though it must measure time, it is actually a measure of distance in space. As you may know, light travels at a speed of about 186,300 miles per second. (This means that light could travel around Earth seven times in one second.) In one year, light travels about six trillion (6,000,000,000,000) miles. The distances to stars and other objects in the universe are measured in light-years because the distances in space are so huge.

Star Distance

The following chart shows five stars that can be seen from Earth and their distances from Earth.

Star	Distance in light-years
Alpha Centauri	4.3*
Sirius	8.7
Vega	27
Arcturus	38
Canopus	230

* As an example, this means that it takes 4.3 years for the light from this star to reach Earth.

Our Galaxy

Imagine living in a city with 200 billion other people. Imagine that if you wanted some milk and bread it would take you almost a million years to travel to the nearest store. If you wanted to visit a friend on the other side of that city it would take more than 100 billion years to get there. That would be quite a large city. But even more amazing, you live in that "city" right now.

The "city" is called the Milky Way. This is the galaxy in which we live, the one in which our solar system is located. (If you'd like to see the Milky Way, you can do so best during the months of July and August. Stand outside where there are no lights—far way from any city is best—and look up. You'll see a "belt" of stars arching across the sky. That "belt" is part of your galaxy, the Milky Way.)

The Milky Way is large, but it's not the biggest galaxy in the universe. If you could see the Milky Way from the side it would look like a flying saucer with a bulge in the middle and thin, flattened sides. If you could see it from the top it would look like an enormous pinwheel, with several "arms" spinning out from the core. The Milky Way is about 100,000 light-years across and 2,000 light-years thick and contains more than 200 billion stars.

Milky Way Galaxy if seen from the side.

Each spiral "arm" of the Milky Way contains gas and dust. These arms are where the youngest stars in the galaxy are concentrated. Many people mistakenly believe that our sun is at the center of this galaxy, but in fact it is about 30,000 light-years away from the center of the Milky Way. Our solar system is located in the Orion (or "local") arm of the Milky Way.

Fantastic Fact

Our sun takes about 220 million years to complete a trip around the center of the Milky Way. Since the "birth" of the universe, the sun has made about 25 circuits of the galaxy.

The Milky Way got its name because it looks like a smudge of milk spread across the sky. It includes all the stars that are visible to

Milky Way Galaxy if seen from the top.

the naked eye, and more. Although we can see about 2,500 stars on a clear summer night, this represents only about one forty-millionth of all the stars in the entire Milky Way.

 ## Hard to See

You'll need:

> sheet of dark-colored construction paper
> white chalk
> tape

What to do:

1. Use the chalk to mark about 30 to 40 white dots in the center of the construction paper. Each dot should be at least one-half inch from every other dot.
2. Tape the construction paper at eye level to the side of your house.
3. Stand about three feet away from the paper and try to distinguish all the dots.
4. Move about six feet away and look at the paper again.
5. Move another six feet away and look at the paper once more.
6. Keep moving backward six feet at a time and look at your paper each time. What do you notice about the dots?

What happens:

When you stand close to the paper you can see each of the individual dots. But each time you move farther away from the paper, the dots become more difficult to see. In fact, the farther away you move, the more the dots seems to "come together" into a big white mass. That's because it's difficult for the human eye to distinguish individual points of light that are close together. This is similar to trying to distinguish the individual points of light from the stars in the Milky Way. Because there are so many of them and because they are so far away, it's hard for us to see most of the individual stars. As a result, the Milky Way appears as a "wash" of stars across the night sky. Also, there is lots of intergalactic dust floating in space, which interferes with our ability to see specific stars, particularly if they are a great distance away.

Galaxies Galore

Typically, galaxies do not exist in isolation; in other words, galaxies usually clump together in clusters held in place by gravity. The cluster of galaxies that includes the Milky Way is known to astronomers as the Local Group. It is a collection of about 30 galaxies spread out over nearly 5 million light-years of space. The most recognizable members (along with the Milky Way) are the Andromeda Galaxy, the Triangulum Galaxy M33, and the Large and Small Magellanic Clouds.

Fantastic Fact

The Andromeda Galaxy is the most distant object visible to the naked eye. Believe it or not, it is 2.3 million light-years away from Earth.

The Local Group is a relatively small collection of galaxies. Two of the largest clusters of galaxies are the Coma Cluster and the Virgo Cluster. Each of these contains thousands of galaxies extending over 20 million light-years of space.

Galaxies come in different sizes and shapes. Following are some of the various configurations:

- *Spiral galaxy*. Like our Milky Way, a spiral galaxy has arms that spin out from its center, similar to a pinwheel. A spiral galaxy's center looks yellowish because lots of older stars are located there.

Spiral galaxy

- *Barred spiral*. Similar to a spiral galaxy, the barred spiral has a bar of stars across its center. The spiral arms spin out from the ends of the bar.

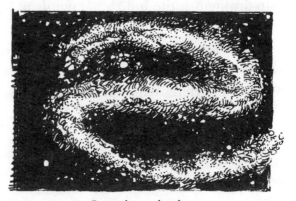

Barred spiral galaxy

Elliptical galaxy

- *Elliptical galaxy.* These galaxies may be small (only a few hundred thousand stars) or extremely large (more than 10 trillion stars). Their shapes may be circular or like a flattened football. They are composed of lots of older stars.

- *Irregular galaxy.* These galaxies have highly irregular shapes and are often quite small. Some are in the shape of a ring, while others have prominent "horns."

Irregular galaxy

The universe is filled with trillions and trillions of objects. Some we see and some we will never see. But despite the existence of all those celestial bodies, it's interesting to remember that about 99 percent of the entire universe is composed of *nothing.* That's right; despite all that you've discovered in this chapter and all you will continue to discover throughout this book, keep in mind that most of space is empty. If, as many scientists believe, the universe is still expanding (remember the Big Bang theory?), that also means that the universe is getting more and more empty every day (that is, the distances between objects are growing all the time).

The universe is an incredible and exciting place, full of discoveries and full of mysteries. In fact, many astronomers will tell you that we know very little about what exists out in space. But this just means that there's that much more to learn. So hop on board for an amazing intergalactic journey to the stars and beyond!

2

Third Rock

Four and a half billion. It's a number almost too large to comprehend. Yet about 4.5 billion years ago our Earth was formed from small rocky objects that collided with one another as they spun around the sun. These collisions generated so much energy that other celestial objects were attracted to and crashed into the developing Earth. After millions and millions of years, Earth reached its current size and began to cool. In the years since those turbulent beginnings, Earth has undergone some remarkable changes. Still, it's amazing to realize that this planet is only a microcosm in the vastness of the universe. It is but one particle in a galaxy of stars, satellites, meteorites, planets, and other celestial bodies.

Many early astronomers believed Earth was a stationary planet, with the other bodies of the universe moving around it. In fact, one of the most respected astronomers in Greece, Claudius Ptolemy, believed that Earth was the center of the universe. This belief lasted for nearly 1,500 years, until a Polish astronomer, Nicolaus Copernicus, suggested that Earth and other planets orbited the sun. This view was argued and debated for many years until the invention of the telescope in the 1600s proved Copernicus right.

With the invention of more sophisticated scientific instruments over the past few centuries, scientists have learned

Earth, as seen from the moon.

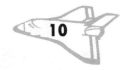

more and more about the planet on which we live. Two of these discoveries are the most significant for understanding Earth's place in the solar system: rotation and revolution.

Rotation

Imagine taking a long piece of wire and sticking it through the center of a Ping-Pong ball so that it sticks out both sides. Imagine having a friend hold the wire at each end while you spin the Ping-Pong ball as fast as you can. What you have created is a rotating object (the Ping-Pong ball spinning around a central point, the wire). This is exactly what Earth does every day. In fact, a *day* is defined as the period of time it takes Earth to spin around its central point (called an axis) once: 24 hours.

Fantastic Fact

Due to the rotational velocity of Earth, a person standing on the equator is moving at a speed of 1,040 miles per hour.

 ## Every Day

You'll need:

> small, round flower pot (with a hole in the bottom)
> > with a couple of inches of soil in it
> two sharpened pencils
> strip of paper (about 8 inches by 1 inch)
> tape

What to do:

1. Tape the strip of paper around the inside rim of the flower pot.
2. Place the flower pot outside on a sunny day (do this early in the morning). Stick the pencil through the hole in the bottom of the pot and through the soil. Make sure the pencil is straight up and down. It should stick out above the top of the soil.
3. At each hour of the day, mark the location of the pencil's shadow on the inside of the pot.

4. At the end of the day (after sunset), take the pot inside and transfer the markings to the strip of paper. You'll notice that the markings move around the rim.

What happens:

The markings on the strip of paper indicate the direction of Earth's rotation (west to east). Earth rotates on its axis. (It spins around an imaginary line drawn through its center). It is this rotation that causes night and day to occur.

As stated above, Earth's rotation is the standard on which the concept of a day (24 hours) is based.

Revolution

Although we do not sense it, we are on a fast-moving spaceship. The spaceship is Earth and it is moving in revolution around our solar system's star, the sun. As Earth revolves around the sun, it moves at about 66,000 miles per hour. The revolution is elliptical (in the shape of a slightly flattened circle). Earth is closest to the sun in January, when it is about 91 million miles away, and farthest from the sun in July, when it is approximately 94 million miles away, with the average distance between the sun and Earth being 93 million miles. It requires 365 days for Earth to make one elliptical revolution around the sun; hence, a year is 365 days long.

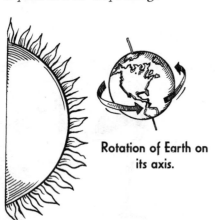

Rotation of Earth on its axis.

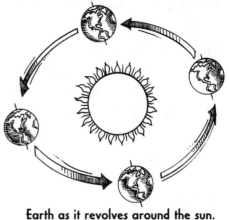

Earth as it revolves around the sun.

Seasons of the Year

As Earth continues its journey around the sun, we experience changes in the duration of sunlight, changes in temperature, and other phenomena. These changes occur primarily because Earth is tilted 23.5 degrees on its axis (that imaginary line that goes through the center of Earth) from "vertical." This tilt causes the Northern

Hemisphere to experience less sunlight during a certain time each year (winter) when Earth is tilted away from the sun. While the Northern Hemisphere is experiencing winter, the Southern Hemisphere (which is then tilted toward the sun) is experiencing summer. This situation is reversed every six months.

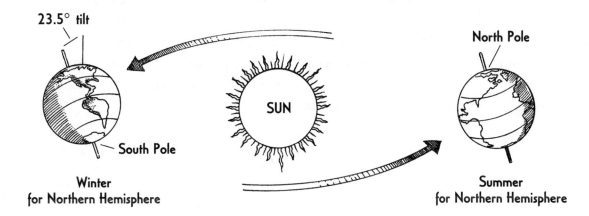

23.5° tilt

South Pole

Winter
for Northern Hemisphere

SUN

North Pole

Summer
for Northern Hemisphere

Solid Earth

The planet Earth consists of four major layers: (1) the inner core, which is basically a solid metallic substance approximately 750 miles thick; (2) an outer core, a thick but mobile liquid about 1,400 miles deep; (3) a mantle, which basically consists of rocks and is about 1,800 miles thick; and (4) the crust, which ranges from 7 to 30 miles in thickness, and is the thin outer layer on which we and all other forms of life depend for our existence.

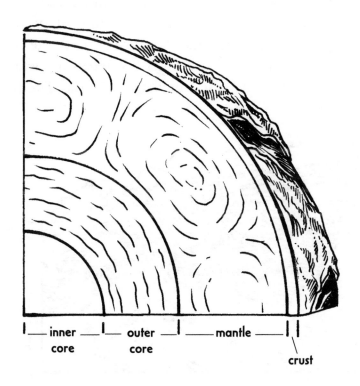

inner core   outer core   mantle   crust

Three basic types of rocks are found on Earth: igneous, sedimentary, and metamorphic.

Igneous Rock

Igneous, fire-formed rock has its origin in Earth's interior (mantle), where it begins as magma, a hot molten substance. The rock forms when the magma cools and crystallizes. If the magma reaches Earth's surface and loses most of its gaseous component, it becomes lava. As magma and lava cool and solidify, they form rocks. Granite, perhaps the best-known igneous rock, solidifies below Earth's surface and has large crystals, a result of a slow cooling process.

Sedimentary Rock

Sedimentary rock is formed by the weathering of bedrock and the movement of the weathered materials by water, gravity, and wind. Sedimentary rocks are formed on Earth's surface. Weathered, tiny particles of rocks, transported by wind, gravity, and water, are eventually deposited and begin to accumulate. As piles of sediments accumulate, the particles at the bottom are compacted by the weight of the upper layers. After a considerable amount of time (millions of years), the sediments are cemented together by minerals that were deposited in the spaces between the pieces of weathered rock. Sandstone is an example of sedimentary rock.

Metamorphic Rock

A rock that has been transformed from a preexisting rock is a metamorphic rock. This type of rock can be formed from igneous, sedimentary, or even other metamorphic rock. The change in classification is caused by extensive heat, pressure, and chemically active fluids, mostly water. Examples of metamorphic rocks are marble (previously limestone, a sedimentary rock) and slate (previously shale, a sedimentary rock).

— sedimentary rock

— metamorphic rock

— igneous rock

Earth's Changing Surface

Although we won't observe a mountain being eroded or a lake being filled with sediment during our lives, the surface of Earth is constantly changing. In many ways our planet is dynamic. Volcanic and earthquake activities are causing the surface to rise in some places, and weathering and erosion are causing some areas to wear down.

Weathering

Weathering is the breaking down of rock at or near Earth's surface. Two of the most common ways in which rocks are broken down are mechanical and chemical weathering.

Through mechanical weathering, large pieces of rock are reduced to smaller pieces with no change in the composition of the rock. Examples of mechanical weathering are freezing, long-term winds, and crushing.

Chemical weathering alters the composition of the rock, creating new materials. (The chemical makeup of the rock changes.) Chemical weathering is caused by water and air, mixed with a small amount of dissolved material, interacting with rocks and causing the internal structure of certain minerals to be altered.

Weathering

Erosion

Many rocks and rock particles are carried from one place to another by wind and glaciers. However, the primary agent of erosion for most rocks is the water in streams and rivers. Water rolls over the surface of rocks for hundreds or thousands of years. This causes a wearing action on the rock, slowly reducing the surface area of the rock. After many years the rock may be worn smooth or may be reduced in size.

Erosion

Bit by Bit, Grain by Grain

You'll need:

white glue
playground sand
tablespoon
water
small coffee can with plastic lid
cookie sheet
medium-size stainless steel bowl

What to do:

1. Mix together six tablespoons of white glue with six tablespoons of sand in a stainless steel bowl.
2. Using the tablespoon, place small lumps of the mixture on a cookie sheet.
3. With adult supervision, place the cookie sheet in an oven heated to 250 degrees F and bake the "rocks" for three to four hours.
4. Remove the rocks from the oven and allow them to cool.
5. Put three or four rocks into a coffee can with some water and fasten the lid securely on top.
6. Shake for four to five minutes and remove the lid.

What happens:

The rocks will have begun to wear down. Some of the rocks will be worn down into sand. The action of the water inside the coffee can causes the rocks to wear against one another. As a result, they break down into smaller and smaller pieces. In nature, this process takes many thousands of years, but the result is the same. Rocks become smaller by being tossed against each other by the force of water. Over time, rocks wear down into sandlike particles that eventually become part of the soil.

Constructive Forces

If weathering and erosion were the only forces affecting our ever-changing world, Earth's surface would be close to a level plane today. However, there are other forces causing changes in its appearance. One is plate tectonics; the other is mountain building.

PLATE TECTONICS

Many scientists believe that Earth's crust is divided into approximately 20 large, rigid, and mobile platelike regions. One of the largest, the Pacific Plate, is located primarily within the Pacific Ocean. Scientists believe that plate tectonics have been active for millions of years and have been a strong influence on the formation of the surface of Earth. These plates move in different directions, causing some of them to collide and others to pull apart. The Pacific Plate, for example, is thought to be moving in a northwest direction, while the North American Plate is moving southeast. The point at which these two plates slide past each other is the San Andreas Fault, one of the most dangerous earthquake regions in the world.

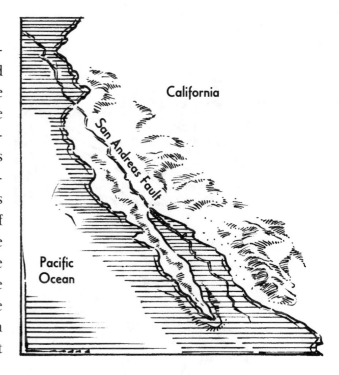

Fantastic Fact

Hawaii and Australia are moving closer to each other by about two and one-half inches each year. At the same time, both are moving away from the South American continent.

MOUNTAIN BUILDING

Although mountain ranges vary in size and appearance, they all have the same basic structure: parallel ridges that consist of a combination of different types of rock. Sometimes there is fossil evidence suggesting that a mountain was once the bed of a lake or other body of water. There are four different types of mountains: (1) folded, (2) volcanic, (3) fault-blocked, and (4) domed.

Folded mountains are formed by great pressure within Earth exerting tremendous sideways force against layers of rocks. This causes a wavelike development, with some of Earth's surface becoming peaks and other areas becoming troughs.

Volcanic mountains are formed by a buildup of lava when volcanoes erupt. They usually take many thousands of years to form.

Fault-blocked mountains are formed by the stress placed on one tectonic plate by another. The stress causes Earth's crust to fracture and sections of the crust to rise in nearly parallel mountain ranges.

Domed mountains are the result of a broad arching of Earth's crust. This occurs because of folding or when magma flows up between two layers of rocks. As the magma accumulates it causes the layers of rocks to move upward and form a large dome.

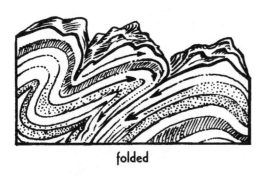

folded

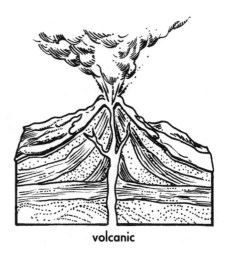

volcanic

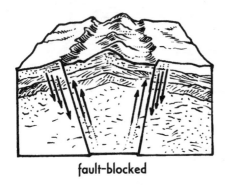

fault-blocked

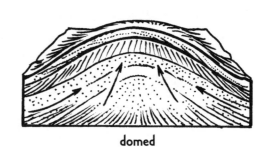

domed

Oceans

They're big! They're huge! They're enormous! They cover more than two-thirds of Earth's surface. They are filled with some of the most amazing, incredible, and fantastic creatures, plants, and geological formations any human being has ever seen. They contain the tallest mountains, the deepest valleys, and the longest mountain ranges on the entire planet. They harbor the deadliest animals and some of the most important lifesaving medicines. They have both the tallest and the smallest plants in the world. The most violent storms and the most beautiful environments are found here. They are home to hundreds of legends, scores of mysterious beasts, and wonderful and marvelous adventures for every person on Earth.

What are they? They're the oceans of the world. Covering nearly 70 percent of Earth's surface, they contain nearly 97 percent of the planet's entire water supply. In fact, of all the planets in our solar system, Earth is the most watery. If you could travel into space (stand on the moon, for example) and look back at Earth, you would be able to see why it's sometimes called "The Water Planet"; indeed, the oceans of the world are visible far out into the reaches of the universe.

Let's take a look at the five major oceans of the world and some of their distinguishing features and characteristics:

Pacific Ocean

- The largest ocean in the world, it covers about one-third of the entire planet.
- It's 63,800,000 square miles in size.
- It reaches almost halfway around the world at its widest point (approximately 11,000 miles wide).
- It's the deepest ocean, with an average depth of 13,215 feet.
- The Mariana Trench, off the Philippines, is the deepest part of any ocean and the deepest point on Earth. It reaches a depth of 35,827 feet (six Empire State Buildings, one on top of the other, would easily fit in the trench).
- It covers an area twice the size of the next largest ocean, the Atlantic.
- It contains more than one-half of all the seawater on Earth, almost as much as the Atlantic and Indian Oceans combined.
- It has more underwater volcanoes (or seamounts) than any other ocean. It also has the greatest number of islands.

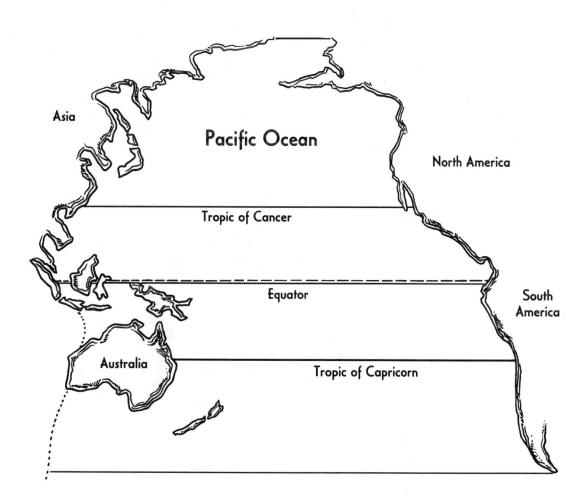

Atlantic Ocean

- It's the second largest ocean, covering an area of about 31,830,000 square miles, or one-fifth of Earth's surface.
- Its widest point is 5,965 miles across.
- It has an average depth of 12,880 feet and is 28,374 feet deep at its deepest point (that's almost as deep as Mt. Everest is tall).
- It contains some of the world's richest fishing grounds. About 90 percent of all the fish caught for food comes from the Atlantic.
- The Mid-Atlantic Ridge, which runs north and south for nearly 7,000 miles, is the longest mountain chain in the world.

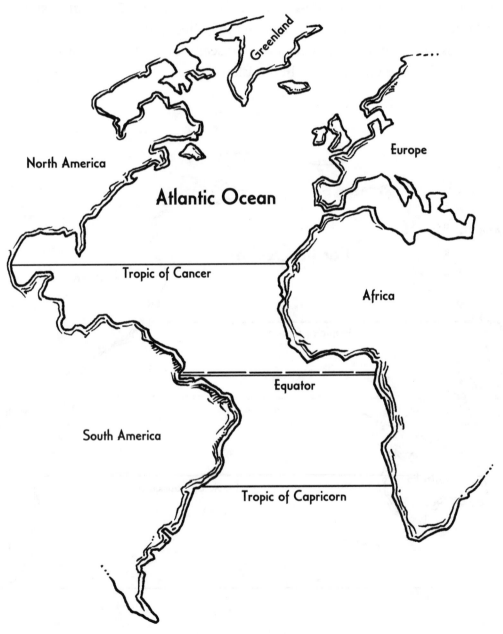

Indian Ocean

- It's the third largest of the world's oceans, covering an area of 28,360,000 square miles.
- It has an average depth of 13,002 feet and its deepest point is 24,441 feet below sea level.
- It contains the saltiest sea (the Red Sea) and the warmest gulf (the Persian Gulf).
- Unlike other oceans, whose currents follow the same path all year long, currents in this ocean change course twice a year.
- Some of the flattest places in the world can be found on the floor of this ocean.

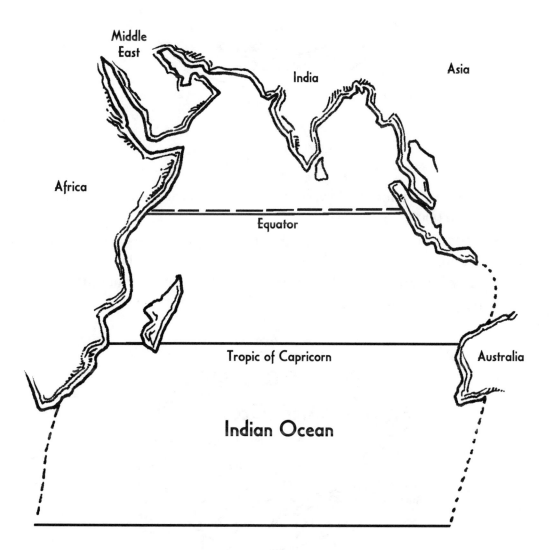

Antarctic Ocean

- The fourth largest of the world's oceans, it covers an area of 13,500,000 square miles.
- In winter, more than half the ocean is covered with ice and icebergs.
- This ocean includes all the waters south of latitude 55° S.

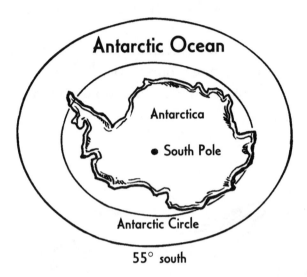

Arctic Ocean

- The smallest and shallowest of the world's oceans, it covers an area of 5,440,000 square miles.
- It has an average depth of 3,953 feet, with a maximum depth of 17,880 feet.
- It's the only ocean almost completely surrounded by land.
- Its waters are covered by sheets of ice, which can be more than 160 feet thick in winter.

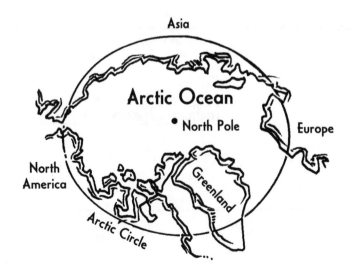

Drip, Drip, Drip

This activity will allow you to create your own fresh water using principles identical to those used by scientists around the world.

You'll need:

water
table salt
measuring cup
measuring spoons
large bowl
small cup
plastic food wrap
small stone

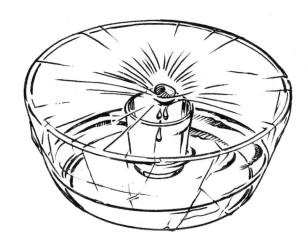

What to do:

1. In the large bowl, mix three teaspoons of salt with two cups of water until the salt is thoroughly dissolved. Use a spoon to taste a small sample of the salt water.
2. Set the small cup inside the bowl, in the middle (see the illustration).
3. Cover the bowl with plastic food wrap.
4. Place the small stone at the center of the plastic wrap (directly over the cup) so that there is a small depression in the plastic wrap.
5. Set this entire apparatus in direct sunlight for several hours.
6. After some time you will notice beads of water forming on the underside of the plastic wrap and dripping into the cup.
7. Afterward, you may want to remove the plastic wrap and carefully taste the water in the cup.

What happens:

The salt water inside the bowl will begin to evaporate into the air inside the bowl. It will condense as beads of water on the underside of the plastic wrap. Because the plastic wrap has a depression at its center, the beads of water will roll down and drip into the cup.

This activity illustrates the process in nature known as solar distillation. It also illustrates the two major components of the water cycle: evaporation and condensation. Distillation involves changing a liquid into a gas (evaporation) and then cooling the gas vapor (condensation) so that it can change back into a liquid. The energy from the sun is able to evaporate water, but not salt, because salt is heavier than water. Thus, the salt remains in the bowl. The water can now be used for drinking purposes and the salt can be used for food seasoning purposes. This process (often referred to as desalination) is used in many countries in the Middle East to make fresh water from salt water.

Earth's Atmosphere

Earth's atmosphere is divided into several layers. The troposphere is the lowest region of the atmosphere; it extends to a height of approximately 11 miles over the equator and 5 miles over the polar region. It is here that we live, play, and work. It is here that our weather develops and occurs. The air we breathe is primarily limited to this area and consists of approximately 78 percent nitrogen, 21 percent oxygen, and 1 percent other gases, including carbon dioxide, helium, and hydrogen.

The second layer of the atmosphere is the stratosphere, which extends from approximately 7 to 31 miles above Earth's surface. This is the region of the ozone

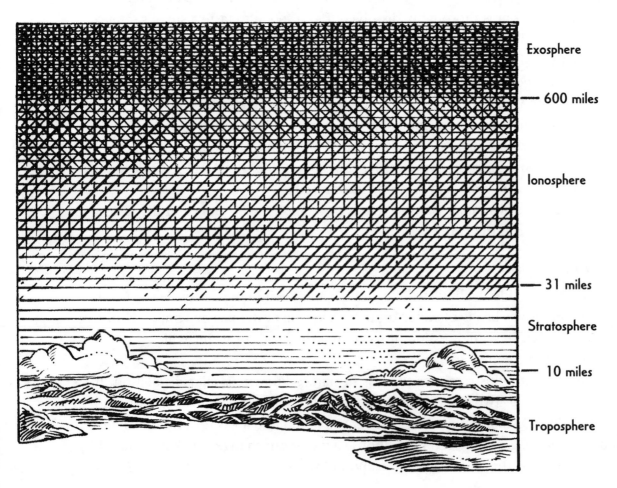

layer. Ozone plays a significant role in sustaining life because it absorbs a considerable amount of Earth's ultraviolet radiation. If this radiation were not filtered, many forms of life would cease to exist.

The ionosphere is an extensive area extending up from about 31 miles above Earth's surface to more than 600 miles. In this area there is a large concentration of electrons and ions, which are formed by the absorption of the sun's ultraviolet radiation and collision with cosmic rays.

The exosphere extends above Earth to the outermost area of the atmosphere. Through the use of artificial satellites, scientists have recognized a large band of radiation around Earth. This area is referred to as the Van Allen Belt.

Wind

Because our planet has an atmosphere, we also have wind. Wind is caused by the unequal heating and cooling of Earth's surface. When air is heated it expands and becomes less dense, or lighter. This causes it to rise, and cooler, denser air moves in to replace it. This movement of air may be a breeze or a strong wind. An excellent example of this occurs at the seashore, where one frequently experiences a sea breeze during the day and a land breeze at night. During the day the land heats much more quickly than the water. The warm air over the land rises and the cool air over the water moves in, creating a sea breeze. At night, the land cools more quickly than the water and the reverse occurs. The warmer air over the water rises and the cooler air over the land moves in to replace it. This phenomenon happens all over the world as a result of the uneven heating of Earth.

Fantastic Fact

One of the strongest winds ever recorded by humans occurred on Mt. Washington, New Hampshire, on April 12, 1934: 231 miles per hour.

Earth File

Equatorial diameter	7,973 miles
Day (time to rotate once)	24 hours
Moons	1
Average orbital speed	66,000 mph
Average distance from sun	93,000,000 miles
Year (time to orbit sun)	365 days
Surface temperature	–22 degrees to 110 degrees F
Earth website	http://www.hawastsoc.org/solar/eng/earth.htm

Our planet is a fascinating place to live, if only because it's the only known planet with life. It's also a fascinating place to learn about, because it's where we live! The more we learn about our world, the more we can begin to appreciate its place in the solar system as well as its place in the universe as a whole.

3

The Sun and the Moon

Everyone, from the most primitive peoples to the most advanced civilizations, is familiar with the sun and the moon. Many peoples and many cultures have worshipped these two celestial bodies and have created gods, statues, and tokens based on beliefs, superstitions, or some degree of scientific accuracy surrounding this star and this satellite.

Scientists also have been fascinated by these two heavenly bodies. Although much of our early knowledge about the sun and the moon was based on conjecture or guesswork, we now know a great deal about these two bodies.

The Sun

The sun is a star. It is also the most important star we know, simply because it is the source of all life on Earth. Green plants need sunlight to grow. Animals eat plants

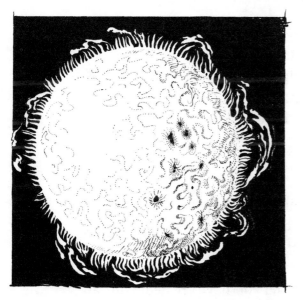

for food, and humans need plants and animals to live. Many of the fuels that we use—coal, gas, oil—are the by-products of once-living things that received sunlight thousands or millions of years ago. Even our daily weather and yearly climate are dependent upon the sun. The sun influences the temperature of the air, the amount of rainfall we experience, the clouds overhead, and the humidity we complain about. In short, the sun is the source of all life on the planet Earth. Without it, we and our plant and animal neighbors would cease to exist.

Sun File	
Age	4,500,000,000 years
Equatorial diameter	864,988 miles
Average distance from Earth	93,000,000 miles
Rotation period at equator	25 Earth days
Surface temperature	11,000 degrees F
Core temperature	27,000,000 degrees F
Sun website	http://www.hawastsoc.org/solar/eng/sun.htm

The sun is at the center of our solar system. It is also the largest body in our solar system. In fact, if it were possible to gather all the planets together, the sun would be 600 times larger than all those planets combined. Even more amazing is the fact that if the sun were hollow it would be able to hold 1,300,000 Earths.

 # Round and Round

This activity will help you appreciate the relative size of the sun.

You'll need:
>   30-foot piece of string
>   measuring tape
>   quarter

What to do:
1. Go to a backyard, park, or playground.
2. Form the string into a circle on the ground. The circle of string should be nine feet in diameter (the distance from edge to edge through the middle of the circle). This circle represents the sun.
3. Place the quarter on the ground along one edge of the string circle. The quarter represents Earth.

What happens:
The equatorial diameter of the sun is 864,988 miles. The equatorial diameter of Earth is approximately 8,000 miles. That means that the diameter of the sun is about 108 times greater than the diameter of Earth.

The string represents the sun and the quarter (which is about one inch in diameter) represents Earth. As you look at the model you have created, you can begin to understand how much bigger the sun is than our own planet as well as every other planet in the solar system.

Many scientists consider the sun to be an ordinary star, with an average size and an average temperature. It has existed for about 5 billion years and is expected to last for another 5 billion years.

All of the sun's energy is generated at its center, where the temperature reaches 27,000,000 degrees F. Here, in a process known as nuclear fusion (in which hydrogen atoms are combined, or "fused," into helium atoms), the sun changes 600 million tons of hydrogen into helium every second. As it does this, it loses 4 million tons of its mass (every second), which escapes as light and heat energy. This is the same process that takes place in a hydrogen bomb, but the sun doesn't explode like a bomb because its gravity holds it together.

The surface of the sun (known as the photosphere) is sometimes "sprinkled" with dark markings known as sunspots. Sunspots are enormous magnetic storms that last for about a week or two. Sunspots usually come and go in a cycle that lasts 11 years. They also give rise to clouds of hot dense gas (known as prominences) that can rise 100,000 miles above the sun's surface.

When a prominence erupts, the fast-moving gas (which may travel at speeds of 1,000,000 miles per hour) travels into space. This "cloud" sometimes hits Earth's atmosphere, causing the air to glow like a neon sign. This results in beautiful auroras (particularly near the North and South Poles), which appear in the night sky as colorful, moving curtains of light. The largest of these solar flares can affect radio communications and may even disrupt electric power stations.

Solar Eclipses

A solar eclipse occurs when the sun (or part of the sun) is blocked from Earth's view by the moon. There are three types of solar eclipses: (1) partial, when only a part of the sun is blocked from Earth's view; (2) total, when the sun is totally blocked from Earth's view; and (3) annular, when the sun's light is still visible around the edge of the moon.

Watch This

Here's how you can safely watch a solar eclipse or "observe" the sun on any given day.

Important Note: You should NEVER look directly at the sun! Because it is so bright, its light can seriously damage your eyes and can even blind you. Astronomers never look directly at the sun, but use special methods and equipment to study the sun safely.

You'll need:

index card
sharp pencil
sheet of white paper

What to do:

1. Using the sharp pencil, poke a hole approximately one-eighth inch in diameter in the middle of the index card. (This may take some practice.)
2. Take the index card and the sheet of white paper outside on a sunny day (or on a day and at a place where a solar eclipse is expected).
3. Stand with your back to the sun. Hold the index card over your shoulder and the sheet of white paper about two feet away (as in the illustration) so that a beam of sunlight goes through the hole in the index card and falls onto the white paper.
4. This will project onto the paper an image of the sun that you can watch safely without injuring your eyes.

What happens:

An image of the sun appears on the white paper. You can make the image brighter by making the hole larger, but the image itself will be fuzzier. You can sharpen the image by using a smaller hole, but the resulting image will not be as bright.

The instruments that scientists use to observe and study the sun operate in much the same way as yours. Scientists, too, are very careful not to look directly into the sun, even with a telescope or other scientific apparatus. The photographs that scientists take of the sun are made through reflecting telescopes that project images of the sun, rather than focusing directly on the sun.

This is a safe way to view the sun without harming your eyes.

Following is a list of solar eclipses due to occur in the next few years and locations where they can best be seen on Earth.

Date	Type	Best viewing area
June 21, 2001	Total	Atlantic Ocean, Southern Africa
December 14, 2001	Annular	Costa Rica, Nicaragua
December 4, 2002	Total	Southern Africa, Indian Ocean, Australia
May 31, 2003	Annular	Scotland, Iceland, Greenland
November 23, 2003	Total	Antarctica
April 8, 2005	Annular	South Pacific Ocean
March 29, 2006	Total	Africa, Asia
August 1, 2008	Total	Greenland, China
July 22, 2009	Total	India, Nepal
July 11, 2010	Total	Chile, Argentina
November 13, 2012	Total	Australia
August 21, 2017	Total	United States
April 8, 2024	Total	United States

If you're interested in learning more about solar (and lunar) eclipses—and seeing them safely from the comfort of your own home—then check out this website: http://www.earthview.com. The site contains tons of educational information about the history, science, and observation of eclipses throughout the world.

The Moon

Perhaps the most familiar sight in the night sky is the moon. From earliest times, humans have always been fascinated with the moon. It is Earth's closest satellite (a satellite is an object that is held in orbit around a larger object, such as a planet). It has been the subject of much observation and scientific investigation, and it is the centerpiece of many traditional stories and legends. Even today, when many people look at the moon they still see an imaginary "man in the moon" formed by markings and craters on its surface.

Many people are surprised to learn that even though we can easily see the moon in the night sky, it does not produce any light of its own. The light we see is actually reflected sunlight. As the moon orbits Earth, its position in the sky changes, so the direction of the sun's light on it also changes. This creates the different "shapes" on the sunlit part of the moon during the course of a month. These changes are known as phases of the moon.

Moon File

Equatorial diameter	2,160 miles
Day (time to rotate once)	27.3 hours
Time to orbit Earth	29.5 days
Average orbital speed	2,112 mph
Average distance from Earth	238,000 miles
Surface temperature (day)	248 degrees F
Surface temperature (night)	–292 degrees F
Moon website	http://www.hawastsoc.org/solar/eng/moon.htm

 # Phase to Phase

To see for yourself why the moon has different phases at different times of the month, you may want to try the following activity.

You'll need:

table lamp (with the shade removed)
basketball

What to do:

1. Place the lamp (without its shade) on a table in the middle of a room. (The lamp represents the sun.) Turn on the lamp and turn off any other lights.

2. Position 1—Face the lamp and stand about 10 feet away from it. (You represent the Earth.) Hold the ball at arm's length in front of you. (The ball represents the moon.) Note that as you look at the ball all the light strikes the back of the ball and there is no light on the front of the ball (the part of the ball facing you). This represents the new moon phase.

3. Position 2—Pivot to the left one-quarter turn as though the moon were orbiting you. Note that you are now able to see light on one-half of the ball's surface (the right-hand portion of the part of the ball facing you). This represents the first quarter phase.

4. Position 3—Turn another quarter turn to your left. (Make sure the ball is not in your shadow; you may have to hold it higher.) Light will now cover the entire surface of the part of the ball facing you. This represents the full moon phase.

5. Position 4—Turn another quarter turn to the left. Light will again be striking one-half of the part of the ball facing you, but this time the left portion. This represents the last quarter phase.

Position 1: new moon

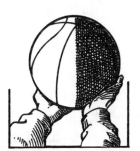

Position 2: quarter moon

Position 4: quarter moon

Position 3: full moon

The basketball is used to demonstrate the four phases of the moon.

What happens:

When the moon (the ball) is on the opposite side of Earth (you) from the sun (the lamp), the moon appears to us to be completely lit (full moon). When the moon is between the sun and Earth, it blocks the sunlight and appears dark (new moon). During the 29.5 days it takes the moon to revolve around Earth, it will have different amounts of sunlight striking its surface. The amount of sunlight that we see reflected from the moon's surface during any month is classified into eight phases (as illustrated on the next page).

Moon Cookies

In this activity you create models of the phases of the moon, models you can eat!

You'll need:

      eight vanilla wafers
      chocolate cake frosting
      dull knife
      the diagram of the eight phases of the moon (see p. 34)

What to do:

1. Place the diagram on a kitchen table.

2. For each of the eight individual phases of the moon, spread chocolate frosting on a separate vanilla wafer, as depicted in the illustration. (***Note:*** The "new moon" will be completely covered with chocolate frosting, while the "full moon" will have no frosting on it.)

What happens:

During each phase of the moon, a portion of the moon's surface is in shadow. The chocolate frosting represents that portion of the moon's surface visible to us that is in shadow. (If you scraped off the frosting, the "moon" would still be there.) It's important to understand that these frosted cookies are just two-dimensional models of a three-dimensional object (the moon).

If you've ever looked at the moon through a telescope, you probably noticed that the surface looked rough and uneven. What you saw were scores of craters, hardened flows of lava, mountain peaks, and even large flat expanses (called "seas," even though there is no water on the moon). Most prominent are the craters, some of which are several miles across. Most of the craters formed more than 3 billion years ago when asteroids and meteoroids crashed into the moon's surface.

Because the moon has no atmosphere, these asteroids and meteoroids didn't burn up as they would have approaching Earth (which does have an atmosphere). Instead, they hit the surface at full speed (from 20,000 to 160,000 miles per hour).

These tremendous collisions generated intense heat that completely destroyed the asteroids in giant explosions while releasing molten lava from inside the moon. The craters (or deep, bowl-shaped holes) that resulted were many times the size of the objects that made them. The lava flows spread across the surface, cooling and settling into large flat areas. This is why the surface looks smooth in some places but pockmarked in other places. Scientists have counted about 30,000 craters on the moon's surface, some as small as one twenty-five-thousandth of an inch across and others as large as 10 miles across.

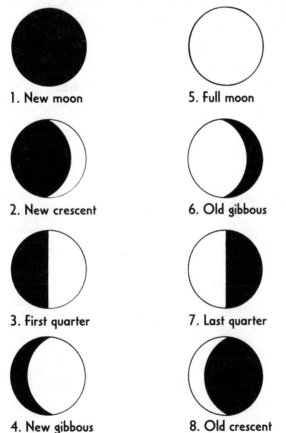

1. New moon
2. New crescent
3. First quarter
4. New gibbous
5. Full moon
6. Old gibbous
7. Last quarter
8. Old crescent

Plop Plop

This (very messy) activity demonstrates how the moon's craters were created.

You'll need:

large baking pan
spatula
sifted all-purpose flour
dried peas, marbles, golf balls

What to do:

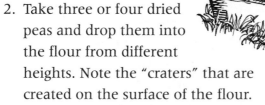
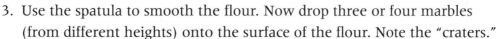

1. Fill the baking pan with about one-half inch of all-purpose flour. Smooth the surface with the spatula. (**Note:** Because this activity will be very messy, I suggest that you do it outdoors on a grassy area.)

2. Take three or four dried peas and drop them into the flour from different heights. Note the "craters" that are created on the surface of the flour.

3. Use the spatula to smooth the flour. Now drop three or four marbles (from different heights) onto the surface of the flour. Note the "craters."

4. Smooth the flour again, and drop three or four golf balls onto the surface. Once again, note the "craters."

5. Smooth the flour (you may need to add more flour) once again. Now drop two or more peas, two or more marbles, and two or more golf balls (all from different heights) onto the surface. What do you notice about the "craters" that form?

What happens:

You noticed that the larger objects made larger craters (and scattered more flour in various directions) than did the smaller objects. The constant bombardment of asteroids on the moon's surface not only created craters but also pulverized the surface, creating a powdery dust very similar to the flour in this activity. In some places this moon dust may range from 3 feet to more than 60 feet in depth. (Imagine standing in a pile of flour that is 60 feet deep!)

Lunar Eclipses

When you are walking on a sunny day, you probably notice that you have a shadow. The shadow results from the fact that you are between the sun and the ground; in other words, you are blocking some of the sunlight from reaching the ground.

The same thing happens in space, too. Earth casts a shadow in space, just as you do on Earth. Whenever Earth is between the sun and some other object, a shadow is projected onto that object. For example, sometimes the moon passes through Earth's shadow. This does not happen very often; because the moon's orbit is tilted slightly, it is not usually lined up with the sun and Earth. However, when it does happen, a full or a partial lunar eclipse occurs. During a lunar eclipse you can see the round shadow of Earth move across the face of the moon and slowly darken it.

Following is a list of lunar eclipses due to occur in the next few years and locations where they can best be seen on Earth.

Date	Type	Best viewing area
January 9, 2001	Total	Asia, Africa, Europe
July 5, 2001	Partial	Australia, Eastern Asia
December 30, 2001	Partial	Asia, Australia, Americas
May 26, 2002	Partial	Asia, Australia, Americas
June 24, 2002	Partial	Africa, Asia, Europe
November 20, 2002	Partial	Europe, Africa, Asia
May 16, 2003	Total	Americas, Africa
November 9, 2003	Total	Americas, Africa, Europe
May 4, 2004	Total	Africa, Europe, Asia
October 28, 2004	Total	Americas, Africa, Europe
April 24, 2005	Partial	Pacific Ocean
October 17, 2005	Partial	Asia, Australia, North America

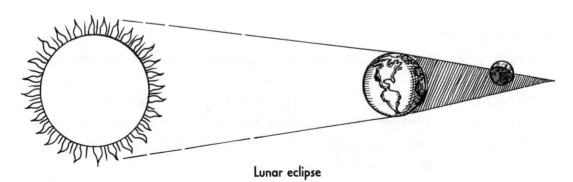

Lunar eclipse

If you are interested in learning more about the moon, there are commercial sources of moon-related items. Following is a list of some of the sources that you can contact (or ask your parents to contact them for you).

MAPS

Earth's moon
National Geographic Society
Educational Services, Dept. 91
Washington, DC 20036
(800) 368-2728

giant map of the moon
Astronomical Society of the Pacific
390 Ashton Ave.
San Francisco, CA 94112
(415) 337-2624

map of the moon
U.S. Geological Survey Map Sales
Box 25286
Denver Federal Center
Denver, CO 80225
(303) 236-7477

GLOBES

Edmund Scientific Co.
101 E. Gloucester Pike
Barrington, NJ 08007
(609) 573-6270

LUNAR PHASE CALENDARS

Celestial Products
P.O. Box 801
Middleburg, VA 22117
(800) 235-3783

EARTH ROCK SAMPLE SETS

Ward's Natural Science Establishment
P.O. Box 92912
Rochester, NY 14692
(800) 962-2660

SLIDES

Glorious Eclipses slide set
Astronomical Society of the Pacific
390 Ashton Ave.
San Francisco, CA 94112
(415) 337-2624

The sun and the moon are the two most prominent celestial bodies for people living on the planet Earth. Not only do they "control" our daily lives, they also provide us with wonderful opportunities to learn more about outer space and our place in the universe.

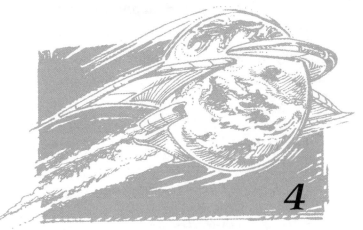

4

Our Solar System

The objects in our solar system and the distances between them are almost unimaginable. Although we know a great deal about the planet we live on, much of what lies beyond our planet remains unknown. How the planets were created, their composition, and the nature of other objects in space are continuing sources of intrigue and mystery. Although we are learning more and more about our solar system, there is still much more to discover, much more to investigate.

You may want to think of the solar system as our neighborhood: a very large neighborhood of one sun and nine planets. But each of our neighbors is unlike every other neighbor: Each has unique features and characteristics totally different from everything else around it.

Our solar system, which is located in the Milky Way galaxy, is believed to have originated approximately 5 billion years ago as a large nebulous cloud, part of the matter and energy expended after the Big Bang. The particles of the clouds were drawn to one another because of gravity; as a result, the cloud began to contract toward its center. During this process the entire cloud spun like a whirl-pool, drawing material together toward the center and building up small concentrations at various distances from the center. These smaller collections of matter eventually became the planets and their moons. You could say that a planet is a big round ball of gas or rock that travels around (orbits) a star (for example, the sun). Some planets are also made of ice and frozen gases.

One of the most fascinating aspects of our solar system is the fact that everything is moving. Planets move around (orbit) the sun. Several planets have moons, which revolve around them. There are lots of celestial bodies such as comets, meteoroids, and asteroids whizzing through space. And the solar system itself moves along on one of the enormous spiral arms of our galaxy.

Spin Around

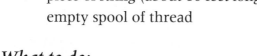

You'll need:

bagel
piece of string (about 10 feet long)
empty spool of thread

What to do:

1. Tie the bagel securely to the end of the string. Slide the other end of the string through the hole in the empty spool.
2. The bagel simulates a planet and the spool simulates the sun.
3. Go outside to a large open area. Hold the spool in one hand (make sure the string passes back and forth easily through the hole) and the end of the string in the other hand. The spool should be perpendicular to the ground.
4. Twirl the bagel in a circular motion around your head so that the string is taut and the bagel whirls around you without touching the ground. Be careful not to hit anyone.
5. As you twirl the bagel, pull the string through the spool so that the distance between the spool and the bagel is shortened.
6. Keep twirling and release some of the string through the spool so that the distance between the spool and the bagel is increased.

What happens:

You may have noticed that when you pulled on the string so that the distance between the spool and the bagel was shortened, there seemed to be more pressure on the string. You also may have noticed that when you extended the string so that the distance between the spool and the bagel was increased, the pressure or force on the string decreased.

The "pull" on the string is similar to the gravitational "pull" of a planet as it revolves around (orbits) the sun. As a planet's distance from the sun (the distance between the spool and the bagel) increased, the gravitational pull toward the sun decreases. With less pull, the orbital speed of the planet decreases too. On the other hand, as the distance between a planet and the sun decreased, the gravitational pull toward the sun increases. With more pull, the orbital speed of the planet increases. That's why planets closer to the sun take less time to revolve around the sun than do planets that are farther away from the sun. Mercury, the closest planet, only takes 88 days to orbit the sun. Pluto, on the other hand, takes more than 248 years to revolve around the sun.

Note: This activity is only a simulation of the revolution of planets in space. Obviously, planets are not attached to the sun by a piece of string.

Mercury

Mercury is named after the Roman god Mercury, who was the swift messenger of the gods. The planet Mercury is swift, too, taking only 88 days to complete one revolution around the sun. It's also our solar system's second smallest planet (after Pluto).

Mercury is only 36 million miles away from the sun. Although that seems like an enormous distance, it's close enough that the temperature on the surface of this planet gets as high as 800 degrees F. Because Mercury is small and so close to the sun, it has almost no atmosphere.

Mercury is pocked with thousands and thousands of craters. These formed during the early stages of the development of our solar system and are due to many collisions with materials from space. In fact, the surface of Mercury is often compared to the surface of our own moon. One of the largest known craters in the universe is located on the surface of Mercury. The Caloris Basin is a central crater measuring more than 800 miles across.

A lot of what we know about Mercury came from the data gathered by the *Mariner 10* space probe, launched in 1973. This spacecraft was able to photograph and map much of the surface of the planet.

Fantastic Fact

After Mercury was formed, it began to cool. As a result, the entire planet started to shrink. Its "skin" was too large for the planet and it began to buckle and wrinkle. The *Mariner 10* space probe was able to photograph most of these wrinkles.

Mercury File

Equatorial diameter	3,031 miles
Day (time to rotate once)	59 Earth days
Moons	0
Average orbital speed	107,132 mph
Average distance from sun	36,000,000 miles
Year (time to orbit sun)	88 days
Surface temperature	800 degrees F (day), –275 degrees F (night)
Mercury website	http://www.hawastsoc.org/solar/eng/mercury.htm

Venus

The second planet in our solar system we'll discuss is Venus. It was named for the Roman goddess of love. Interestingly, all the physical features on the surface of this planet have been named for real or imaginary women.

Venus has often been referred to as our sister planet because it shares some of the characteristics of the planet Earth. Venus is often seen as a bright object in the night sky, second in brilliance to the moon. It differs from Earth in one significant aspect: It rotates from east to west, while most planets, including Earth, rotate from west to east. This means that on Venus the sun would rise in the west and set in the east.

Although Venus is farther from the sun than Mercury, it is actually hotter on the surface. This is because Venus is covered by a layer of sulfuric acid clouds. As a result, much of the heat that reaches Venus gets trapped beneath those clouds and can't escape. It's almost as though the surface of this planet has been turned into a giant greenhouse. In fact, the surface frequently gets more than four times as hot as boiling water.

Fantastic Fact

Venus takes 243 Earth days to make one rotation and 225 Earth days to make one revolution around the sun. Thus, on Venus, a day is actually longer than a year.

Venus File

Equatorial diameter	7,521 miles
Day (time to rotate once)	243 Earth days
Moons	0
Average orbital speed	78,364 mph
Average distance from sun	67,200,000 miles
Year (time to orbit sun)	225 days
Surface temperature	870 degrees F
Venus website	http://www.hawastsoc.org/solar/eng/venus.htm

41

Mars

Mars is named after the Roman god of war. In the night sky there is a reddish hue to its surface and the atmosphere surrounding it. In fact, it is frequently referred to as the "Red Planet." For centuries it has been one of the most intriguing planets in our solar system.

Mars has a very thin atmosphere made primarily of carbon dioxide. It's also a very dry planet, with no liquid water. However, many scientists now believe that Mars may have had water at one time. This theory has been supported by evidence gathered by the *Pathfinder* mission to the surface of Mars in 1997. This mission placed a small robot rover, known as "Sojourner," on the planet's surface, where it traveled for several weeks, gathering important information.

Despite what some science fiction stories would have us believe, Mars is considered to be a lifeless planet. Enormous extinct volcanoes are spread across its surface, and fierce martian winds whip up clouds of dust that sometimes cover the entire planet. Lots of craters are sprinkled across the landscape of Mars, caused by collisions with materials from space over many millions of years. Mars, like Earth, has two poles, each of which is covered by white icy caps. These ice caps frequently enlarge and recede, depending upon the martian seasons.

Fantastic Fact

The largest volcano in the entire solar system is on Mars. Olympus Mons is as wide as the entire state of Arizona and is more than 17 miles high. (Mt. Everest is only 5 miles high.)

Mars File

Equatorial diameter	4,217 miles
Day (time to rotate once)	24.6 hours
Moons	2
Average orbital speed	53,980 mph
Average distance from sun	141,600,000 miles
Year (time to orbit sun)	687 days
Surface temperature	700 degrees F (day), −207 degrees F (night)
Mars website	http://www.hawastsoc.org/solar/eng/mars.htm

Jupiter

Jupiter is the largest planet in the solar system. It is 300 times heavier than Earth and more than twice as heavy as all the other planets *combined*. Jupiter is made up primarily of the gases hydrogen and helium. Its atmosphere, on the other hand, is composed of other gases such as methane and ammonia. This combination of gases often changes color, frequently making the planet look different (through a telescope) over a period of several nights.

One of the most distinctive features of this planet is the Giant Red Spot, which has been observed on the surface of the planet for at least 300 years. This swirling whirlpool of gases is an enormous storm cloud that seems to rage constantly on the planet. It has been estimated that this storm is about 25,000 miles long and 7,000 miles wide and that its winds may be blowing at up to 900 miles per hour.

Jupiter has at least 16 moons, 4 of which have been studied by astronomers since the seventeenth century. Ganymede is the largest moon; in fact, it is larger than the planet Mercury and is also the largest moon in the entire solar system. Europa is the smallest of these four moons and its surface is similar to that of a billiard ball: round and smooth. Callisto is pocked with thousands of craters across its surface and is considerably bigger than our own moon. The last of these four primary moons is Io, which is one of only two moons in the solar system with active volcanoes.

Jupiter File

Equatorial diameter	88,730 miles
Day (time to rotate once)	9.8 hours
Moons	16
Average orbital speed	29,216 mph
Average distance from sun	483,400,000 miles
Year (time to orbit sun)	11.9 Earth years
Surface temperature	–238 degrees F
Jupiter website	http://www.hawastsoc.org/solar/eng/jupiter.htm

Bigger Than Big

You'll need:

three sheets of paper
ruler
pencil
scissors
cellophane tape

What to do:

1. On one sheet of paper, draw a circle that is one inch in diameter. Cut out the circle. Set this circle aside.
2. Use the cellophane tape to tape the other two sheets of paper together along the long edge of each sheet, forming a large rectangle.
3. On the joined pieces of paper, draw a circle that is 11 inches in diameter. Cut out the circle.
4. Place the small circle inside the large circle.

What happens:

The small circle represents the planet Earth. The large circle represents the planet Jupiter. Notice the difference in size between these two "planets." If you want a challenge, try to figure out how many paper "planet Earths" can fit inside the paper "planet Jupiter."

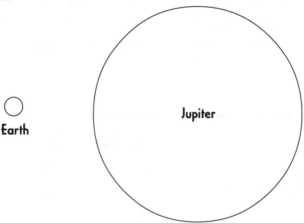

Fantastic Fact

Jupiter is more than twice as heavy as all the other planets of the solar system combined.

Saturn

Saturn is the second largest planet in the solar system and, arguably, the one that has been observed most frequently by generations of astronomers. This planet is noted for its rings, although it is not the only planet with rings. What makes Saturn's rings different is that they are distinctive and can be seen clearly from Earth with powerful telescopes. Each of the rings is made up of bands of ice and rocks formed long ago, perhaps when an asteroid came too close to Saturn and was torn apart by the strong gravity of the planet. The rings, made of billions of chunks of material, may be up to 170,000 miles wide and only 1 mile thick. Most of the chunks of material in each ring are about the size of a baseball, although there are a few the size of a small house. Although we can see three rings from Earth, Saturn is actually composed of tens of thousands of rings.

Saturn has more moons than any other planet: 18 at last count. Many scientists believe that there are still some moons yet to be discovered. Saturn's largest moon, Titan, is bigger than Mercury and is the only moon in the solar system with an atmosphere. This atmosphere is composed primarily of nitrogen, while the surface is liquid methane.

Fantastic Fact

Saturn, with a density of less than half an ounce per cubic inch, would float like a bar of soap if you could find a bathtub big enough.

Saturn File

Equatorial diameter	74,600 miles
Day (time to rotate once)	10.2 hours
Moons	18
Average orbital speed	21,565 mph
Average distance from sun	886,700,000 miles
Year (time to orbit sun)	29.5 Earth years
Surface temperature	−290 degrees F
Saturn website	http://www.hawastsoc.org/solar/eng/saturn.htm

Uranus

Like Jupiter and Saturn, Uranus has no solid surface. Its outer layers are made of hydrogen, helium, and methane. It is this methane that gives Uranus a slightly blue color when viewed through a telescope.

The most distinctive characteristic of this planet is that it is lying on its side. That's right, Uranus does not rotate side to side like Earth, but rather top to bottom. For example, the planet Earth is tilted about 23.5 degrees on its axis from "vertical." It is this tilt that causes the seasons. Uranus, on the other hand, is tilted at an angle of 98 degrees, which means it spins on its side.

Here's how you can experience this "tilt" yourself: Stand straight up. Your "axis" is now 0 degrees. Next, lie down on the ground. Now your axis is on its side, or at 90 degrees. By lying down on your side, you are tilted at approximately the same angle as the planet Uranus.

This extreme tilt in the planet's axis also means that its poles face the sun (for 42 years at a time). Because they get so much light, the polar regions are warmer than the area around Earth's equator.

Uranus has 15 known moons, most of which are quite small, less than 90 miles in diameter. Its largest moon, Ariel, is 720 miles in diameter. Much of what we know about this planet was relayed back from the spacecraft *Voyager 2*, which flew by Uranus in 1986 after a journey of nearly eight and one-half years. It was this spacecraft that identified four new rings encircling the planet (Uranus has 13 rings in all).

Uranus File

Equatorial diameter	31,600 miles
Day (time to rotate once)	17.2 hours
Moons	15
Average orbital speed	15,234 mph
Average distance from sun	1,784,000,000 miles
Year (time to orbit sun)	84.1 Earth years
Surface temperature	–346 degrees F
Uranus website	http://www.hawastsoc.org/solar/eng/uranus.htm

Fantastic Fact

The planet Uranus is named after the Greek god of the sky. It was discovered by the English astronomer William Herschel in 1781. Interestingly, the original name proposed for this planet was "Herschel."

Neptune

Neptune was discovered in 1846 as a result of complicated mathematical calculations by astronomers in France, England, and Germany. Slightly smaller than Uranus, it has eight moons. The two largest moons are Triton and Nereid. Triton is the coldest object in the solar system, with temperatures plunging to an icy –455 degrees F. The surface of Triton has melted and refrozen repeatedly since its creation, forming a network of huge cracks over its surface. It also does something very unusual for a moon: It orbits the planet in the opposite direction from the planet's own spin.

Neptune, like Uranus, has an atmosphere of hydrogen, helium, and methane. Often a large dark spot will appear on the surface of this planet, but unlike Jupiter's spot, Neptune's spot will come and go at irregular times. The surface of the planet is violent, with winds greater than 600 miles per hour, much greater than any hurricane on Earth.

Fantastic Fact

Aside from the planet Earth, Neptune is the bluest planet in the solar system. It was named for the Roman god of the sea.

Neptune File

Equatorial diameter	30,700 miles
Day (time to rotate once)	19.4 hours
Moons	8
Average orbital speed	12,147 mph
Average distance from sun	2,794,400,000 miles
Year (time to orbit sun)	165 Earth years
Surface temperature	–360 degrees F
Neptune website	http://www.hawastsoc.org/solar/eng/neptune.htm

Pluto

Pluto was named for the Greek god of the underworld. Not a great deal is known about this mysterious frozen rock, the smallest known planet in our solar system. Many people say that Pluto is the farthest planet from the sun. That is only partially correct. Pluto has a strangely elongated orbit around the sun. It spends 20 of the 248 years it takes to revolve around the sun inside the orbit of Neptune. (This occurred most recently between 1979 and 1999.) For those 20 years Neptune was the most distant planet in our solar system.

For most of its very long year the materials that make up Pluto's surface are frozen. But when Pluto moves closer to the sun (within about 3,000,000,000 miles), the solid materials turn into gases and the planet has a thin atmosphere of methane, nitrogen, and carbon dioxide. As it moves to its farthest point from the sun (5,000,000,000 miles), these materials freeze once again, leaving the planet completely frozen.

Pluto has one known moon, Charon. Charon orbits the planet in six days and nine hours. This is the same amount of time that it takes Pluto to rotate once. This means that Charon always stays above the same point on the surface of the planet.

Fantastic Fact

Pluto is the only planet to have been discovered by an American. It was first sighted by a 24-year-old astronomer named Clyde Tombaugh in February 1930.

Pluto File

Equatorial diameter	1,420 miles
Day (time to rotate once)	6.4 Earth days
Moons	1
Average orbital speed	10,604 mph
Average distance from sun	3,656,000,000 miles
Year (time to orbit sun)	248 Earth years
Surface temperature	–360 degrees F
Pluto website	http://www.hawastsoc.org/solar/eng/pluto.htm

 # Far, Far Away

You'll need:

10 index cards
marking pen
10 Popsicle sticks (available at any hobby or craft store)
white glue
planet materials (see below)
measuring tape
football field

What to do:

1. Using the marker, write the name of each planet (and the "sun") on a separate index card.
2. Glue each card to the top of a Popsicle stick.
3. Have an adult take you to the nearest high school football field.
4. Push the stick with the "sun" index card on it into the ground at one of the end zone lines on the field.
5. Use the measuring tape to measure the following distances on the football field (you may also wish to use the yard lines on the football field). At each measurement, push the appropriate planet "sign" into the ground.

Planet	Distance from "sun"
Mercury	2.5 feet
Venus	4.5 feet
Earth	6.5 feet
Mars	10 feet
Jupiter	11.5 yards
Saturn	20.5 yards
Uranus	41.5 yards
Neptune	65 yards
Pluto	86 yards

6. After you have placed all the signs on the field, return to each one and place beside it the corresponding object listed in the next chart.

Planet	Corresponding "size"
Mercury	Pea
Venus	Walnut
Earth	Ping-Pong ball
Mars	Bean
Jupiter	Soccer ball
Saturn	Head of lettuce
Uranus	Grapefruit
Neptune	Baseball
Pluto	Peppercorn

What happens:

The way the "planets" are spread out over the football field illustrates the enormous distances that exist in our solar system. The distances between the planets and between each planet and the sun are staggering. This model simulates (on a much smaller scale) the incredible distances between celestial bodies.

The relative sizes of the planets to one another are also illustrated with this model. Jupiter (soccer ball), the largest planet, is considerably bigger than Pluto (peppercorn), the smallest planet.

It is important to remember that this simulation is not entirely accurate because in reality the planets do not string out in a straight line from the sun. As you know, the sun is at the center of our solar system, but the planets are "arranged" in various circular or elliptical orbits around it. You would need a tremendously large field or open area to depict this arrangement accurately.

The following chart illustrates the enormous distances in space. Remember that light travels at a speed of approximately 186,000 miles per second. (That's about 11,160,000 miles per minute or 669,600,000 miles per hour.) This chart illustrates how long it takes a beam of light to travel from the sun to each of the nine planets.

Planet	Time / light from sun
Mercury	3 minutes
Venus	6 minutes
Earth	8 minutes
Mars	13 minutes
Jupiter	43 minutes
Saturn	81 minutes
Uranus	162 minutes
Neptune	254 minutes
Pluto	332 minutes

Wow! Our solar system is truly amazing. We know a lot about it, but there is still much to be discovered. Space probes, satellites, and powerful telescopes are helping us learn more about this magical, marvelous system.

5 Stars and Constellations

Stars

Believe it or not, stars go through a life cycle just like plants and animals. Of course, stars are not living objects like a pine tree or an elephant, but they do go through a regular pattern or cycle of events, from their birth to their death.

First of all, it's important to understand what a star is. Stars are formed when clouds of gas and dust (called a nebula) get squeezed together by gravity. This squeezing motion causes the gas and dust molecules to rub against one another, producing a lot of heat. This ball of gas (usually hydrogen) acts just like an enormous nuclear power plant, producing tremendous amounts of heat and energy. The star is now called a protostar. In fact, the temperature at the center of this new star may reach 18,000,000 degrees F.

The enormous heat inside the star changes the hydrogen atoms into helium atoms. When this happens there is an "atomic reaction" and a flash of energy is given off. Billions of these flashes of energy are what make a star visible, particularly at night, when they can be seen against the backdrop of a darkened sky. It's important to remember that this "birth" takes place over many millions of years.

Hydrogen is the "fuel" for every star. The rate at which a star burns hydrogen determines how long it will "live." Stars that burn their hydrogen very quickly are the hottest stars and are blue-white in color. Stars that burn their hydrogen steadily, like our sun, are yellow in color. Stars that burn hydrogen very slowly are the coolest stars and are red in color.

There are several different categories of stars; they are classified according to the temperature at the surface. Following are the types of stars found throughout our universe.

Type of star	Temperature
Blue stars	36,000 – 90,000 degrees F
Blue-white stars	18,000 – 36,000 degrees F
White stars	14,000 – 18,000 degrees F
Yellow stars	10,000 – 14,000 degrees F
Orange stars	9,000 degrees F
Red stars	7,000 degrees F

Fantastic Fact

Our sun (a star) was formed about 5 billion years ago. It is expected to exist for another 5 billion years.

One question that many people ask is: Why can't I see stars in the daytime? Another question is: Do stars disappear during the day? The following activity will help you answer those questions.

 # Twinkle, Twinkle, Little Star

You'll need:

>3-by-5-inch index card
>paper punch
>two sheets of typing paper
>flashlight
>cellophane tape

What to do:

1. Use the paper punch to punch five to seven holes randomly on the index card.
2. Tape the index card to the middle of one sheet of paper.
3. Tape the other sheet of paper on top of the first (so that the index card is "sandwiched" between the two sheets of paper).
4. Hold your paper "sandwich" in front of you. Turn on the flashlight and aim it at the front of the "sandwich" (at the spot where the index card is positioned).
5. Now take the flashlight and hold it behind the paper "sandwich," aiming it at the spot where the index card is.

What happens:

When you hold the flashlight at the front of the paper "sandwich," you aren't able to see the holes in the index card. This is because the light is bright and is coming from the same direction your eyes are looking from. This is similar to daylight hours, when the stars are in the sky but the brilliant light of the sun is coming from behind you. As a result, you can't see the stars.

When you place the flashlight behind the paper "sandwich," the light comes through the holes in the index card, making them easy to see. This is similar to what happens at night. The stars give off their own light and that light is not influenced by the light from the sun (which has set).

So, stars are always there. It's just because of the sunlight during the day that we cannot see them. When there is no sunlight and the only light comes from the stars themselves, then we can see them.

Fantastic Fact

The light we see from the stars may have begun its journey to Earth hundreds of years ago. Because of the vast distances in our universe, the starlight seen today may have been created a long time ago and is just now reaching us after its incredibly long journey.

After many billions of years, stars begin to die. This happens when they run out of fuel. The way a star dies depends upon how big it is. A medium-sized or smaller star may live for billions of years. When its central core runs out of hydrogen gas, the core becomes hotter and swells up. This swollen star is known as a red giant. Eventually, the star collapses in on itself, leaving the dead core, known as a white dwarf. A white dwarf has no fuel and its heat slowly leaks away into space. As it cools, it changes color from white to yellow to orange to red (see the chart on p. 52). It finally finishes as a cold, black ball.

Large or massive stars have a much more spectacular death. Interestingly, large stars do not live as long as small stars: They may only last for a few million years. They usually die in a dramatic explosion, scattering their innards into space.

These exploding stars are called supernovas. What is left is a very small, extremely dense body known as a neutron star. Some neutron stars are called pulsars. That's because they give off pulses of light and radio waves as they spin. If the core is super massive, gravity overwhelms it and it turns into a black hole. The gravity in a black hole is so strong that even light from a few miles around cannot escape. Because no light can be seen, black holes are invisible.

Kinds of Stars

Stars are differentiated by their colors. They can also be classified by how they behave. Basically, there are two types of stars: variable and binary.

- *Variable stars.* Many stars have a regular cycle of fading and getting brighter. These are known as variable stars. Often these cycles are predictable. For example, the star called Delta Cephei goes from bright to dim to bright in a cycle lasting five days and nine hours.

- *Binary stars.* Binary stars are pairs of stars that orbit each other. They are able to do this because of the force of each other's gravity. Sixty percent of the stars in the universe are binary stars. Sometimes the stars will differ in color and brightness. Sometimes they may be close to each other (called contact binaries) or great distances apart. When a binary system star blazes very brightly it is called a nova.

 Spin Cycle

You'll need:

> sheet of black construction paper
> white chalk
> sharpened pencil
> ruler
> masking tape
> scissors

What to do:

1. Cut a circle from the construction paper that is about five inches in diameter.
2. Use the piece of chalk to mark two dots on the paper opposite each other on an imaginary line drawn around the middle of the circle (see the illustration).

3. Insert the sharpened pencil through the middle of the paper circle and use a strip of tape to secure the pencil to the bottom of the paper.
4. Place the pencil between your hands and twirl it back and forth.

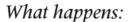

What happens:

You'll note that when you twirl the pencil back and forth the two white dots seem to "blend" together. In fact, it may seem as though there are rings appearing on the circle, particularly if you twirl it fast enough. Something similar happens with binary stars. The speed of the orbit of two or three stars around one another makes it seem as though there is one star instead of two or three. So, even if two or three stars are thousands of miles apart, their rotational speed is so rapid that they become one source of light. You may want to experiment with this by spinning the pencil quickly and then spinning it slowly to see what happens to the two dots. Add a third dot and see if anything changes.

Star Brightness

The brightness of a star is called its magnitude. The Greeks invented a scale for measuring magnitude more than 2,000 years ago; it is still in use today, although in a somewhat modified form. The higher the number, the dimmer the star. Stars with a magnitude of 6 will be very dim in the night sky. Magnitude 6 stars are those we can see with the naked eye. With binoculars we can see magnitude 9 stars, and with a small telescope we can see stars with a magnitude of 12 or 13.

Listed below are the five brightest stars that can be seen from Earth:

Star	Magnitude	Light-years from Earth
Sirius	1.5	8.7
Canopus	0.7	230
Alpha Centauri	0.3	4.3
Arcturus	0.06	38
Vega	0.04	27

Constellations

Humans have always been fascinated by the stars above. Often when observing the heavens human beings would imagine patterns or drawings within groups or clusters of stars. Many cultures around the world have invented their own unique and distinctive constellations. But it is those invented and described by the ancient Babylonians and Greeks more than 2,500 years ago that we know today.

A constellation can be described as a pattern of stars in the night sky. What is important to remember is that the stars that make up a particular constellation are not really close to one another. Often stars that look close to one another are millions of miles apart—they just happen to form a two-dimensional pattern when seen from Earth. For example, the chart below lists the eight stars in the constellation Orion (The Hunter). Notice how many light-years away from Earth each of the stars is.

Star	Light-years from Earth
Meissa	320
Betelgeuse	1,360
Bellatrix	1,660
Alnitak	400
Alnilam	160
Mintaka	380
Saiph	40
Rigel	340

The Big Dipper and the North Star

One of the most well known constellations is the Big Dipper. Why is it called the Big Dipper? Simply because it looks like a giant dipper or ladle for water.

Here's how you can locate this constellation: On a cloud-less, moonless night away from the bright lights of a city, face north (use a compass) and look up. You'll see a pattern of seven stars (see the illustration), four of which form the "bowl" of the Big Dipper and three of which form the "handle." Once you've sighted the Big Dipper, draw an imaginary line along the two stars on the outer end of its bowl. That line will take you to a star on the handle end of the Little Dipper constellation.

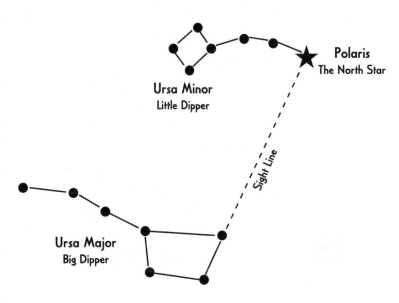

Ursa Minor
Little Dipper

Polaris
The North Star

Sight Line

Ursa Major
Big Dipper

The star on the handle end of the Little Dipper is one of the most famous in the sky: the North Star, or Polaris. For centuries sailors have used the North Star to guide them across endless expanses of ocean. The North Star and the Big Dipper also played an important role in American history.

During the years before the Civil War, large numbers of black slaves tried to make their way to the free northern states and Canada. They traveled along an imaginary route known as the Underground Railroad. Because they didn't have signposts or maps to guide them, the runaway slaves often relied on the stars to point them in the right direction. A simple song popular at that time, "Follow the Drinking Gourd," contained helpful clues for a successful journey:

When the sun comes back
And the first quail calls
Follow the Drinking Gourd.
For the old man is a-waiting for to carry you to freedom
If you follow the Drinking Gourd.

The riverbank makes a very good road.
The dead trees will show you the way.
Left foot, peg foot, traveling on,
Follow the Drinking Gourd.

The river ends between two hills
Follow the Drinking Gourd.
There's another river on the other side
Follow the Drinking Gourd.

When the great big river meets the little river
Follow the Drinking Gourd.
For the old man is a-waiting for to carry you to freedom
If you follow the Drinking Gourd.

This song has several clever hints hidden in its lyrics that helped the slaves find their way north. For example, the "Drinking Gourd" is the Big Dipper, which, of course, helped to locate the North Star. In the first verse, "When the sun comes back / And the first quail calls" are code words for spring, which was the best time of the year to travel north. The river that "ends between two hills" in the third verse is the Tombigbee River in Mississippi. And the "great big river" in the fourth verse is the Ohio River. With these "directions" slaves were able to flee from the South through the North and eventually into Canada.

Following is a nifty device you can use to help locate yourself anywhere in the Northern Hemisphere.

 # Find Me!

You'll need:

    3-by-5-inch index card
    pencil
    protractor
    piece of light-colored string or thread (about 7 inches long)
    small washer

What to do:

1. Make a small hole in one corner of the index card using the point of the pencil.
2. Using the hole as a center, draw a right angle, as illustrated in the following diagram.
3. Use a protractor to draw line segments that divide the right angle into 10-degree angles, as shown.

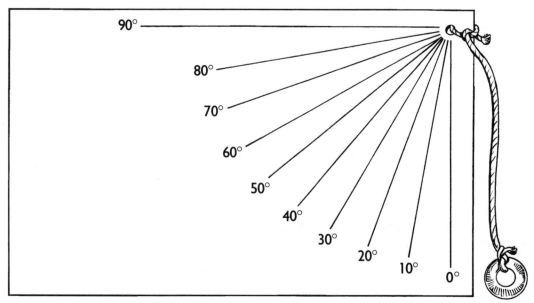

4. Tie one end of the piece of string or thread to the hole in the corner of the index card.

5. Tie the other end of the string to the small washer.

6. To use this "Latitude Finder," locate the North Star (Polaris). Locate the Big Dipper in the northern part of the sky. The Big Dipper is a constellation of seven stars. Three stars form the "handle" and four stars form the "bowl." Find the two stars on the outer end of the Big Dipper's bowl. Use your finger to follow these two stars directly up to a star of average brightness. This is Polaris, the North Star.

7. Hold the Finder and place one corner of the Finder near your eye. (See illustration below.)

8. Aim the opposite corner of the Finder (the corner nearest to the hole) directly at the North Star.

9. Allow the washer to hang freely.

What happens:

When you have Polaris sighted at the end of the index card, the string will hang toward Earth. The string will fall against the right angle you have drawn on the index card. The line along which it falls roughly indicates the latitude of your position on Earth (in the Northern Hemisphere).

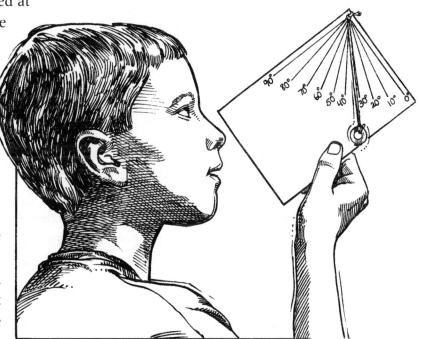

This works because the North Star remains stationary, almost directly over the North Pole. If you were standing at the North Pole and sighted the North Star with the Finder, it would indicate that you were at latitude 90° N. On the other hand, if you stood directly on the equator and sighted the North Star, your Finder would indicate that you were at latitude 0°, or right on the equator.

This Finder plots a rough position of where you are on Earth. You may want to check a map or atlas to confirm your exact location. Following are some latitudes of selected major cities in the United States you can use for comparison:

City	Latitude
Los Angeles	34.03° N
San Francisco	37.45° N
Seattle	47.36° N
Denver	39.44° N
Phoenix	33.30° N
Dallas	32.45° N
Chicago	41.49° N
New Orleans	30.00° N
Miami	25.45° N
Philadelphia	40.00° N
New York	40.40° N
Boston	42.15° N

On pages 61–63 are pictures of 12 of the most well known constellations in the Northern Hemisphere. These are only a few of the 88 known constellations. You may want to use these illustrations to help you locate constellations whenever you look up at the night sky. For further exploration, plan to purchase a star map at your nearest bookstore or teacher supply store.

Constellation Websites

Following are three fantastic websites at which you can obtain up-to-the-minute information about stars and constellations.

- **http://einstein.stcloudstate.edu/dome/clicks/constlist.html**

This website will provide you with close-up photographs and illustrations of most of the major constellations. This is a great starting place for any nighttime exploration of the skies.

- **http://www.geocities.com/capecanaveral/launchpad/1364/ constellations.html**

This site has lots of illustrations and wonderful descriptions of most of the constellations. Super graphics highlight this site, along with very descriptive narratives.

- **http://www.astro.wisc.edu/~dolan/constellations/ constellations.html**

Tons and tons of information about the constellations can be found at this site, one of the most popular on the Web. During a two-year period it was accessed over 2 million times, so you know it's good!

Taurus

Canis Major

Cancer

Draco

Leo

Hercules

Lyra

Pegasus

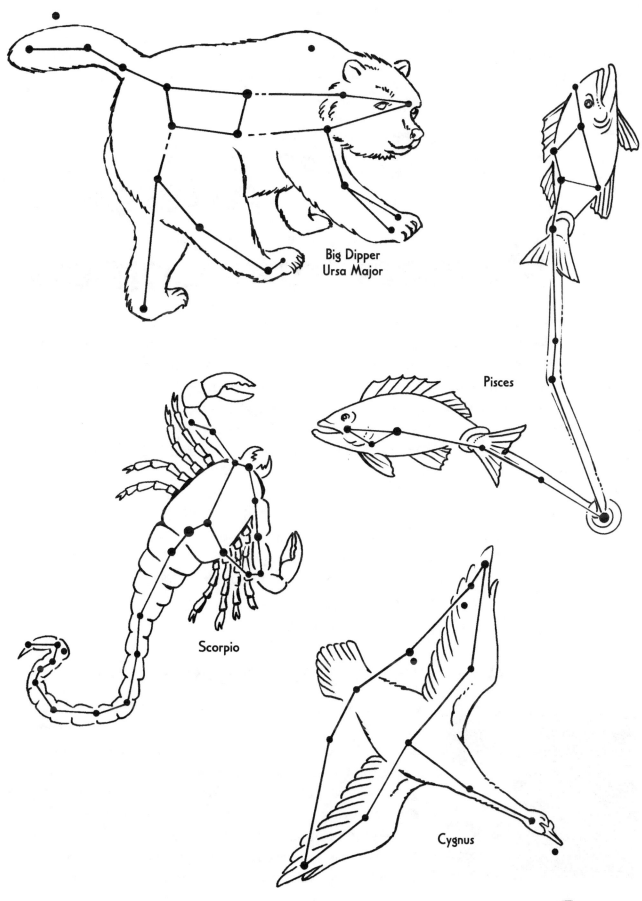

Big Dipper
Ursa Major

Pisces

Scorpio

Cygnus

6

Asteroids, Meteors, and Comets

Have you ever been outside on a warm summer night or early fall evening and looked up at the sky? What did you see? Did you see an array of stars strung out across the night sky? Did you see a planet? Did you see the moon high overhead? Or perhaps you saw something or several things moving across the night sky, streaking from one edge to the other. If you were with someone, there were probably some gasps or ohhhhhs and ahhhhhs. Somebody may have said that the object was a shooting star or a falling star: a bright point of light rapidly moving across the sky with a long, arcing tail.

Interestingly, although people for thousands of years have been fascinated by the objects often referred to as "shooting stars," there really isn't any such thing as a star that shoots across the sky. As you recall from Chapter 5, stars are stationary: They are fixed points of light in the sky (relative to Earth). And, most important, they don't "fall" from the sky. Although we often refer to these objects as "shooting stars," they are really meteors.

Asteroids and Meteoroids

As you know by now, the solar system has many different kinds of celestial bodies. Some are large (planets) and some are small (dust). Most of these objects were created during the Big Bang. Some of the smaller objects, known as asteroids, are actually small rocky bodies of material that never came together in the early days of the solar system to form planets. Most of the asteroids in our solar system lie in an asteroid belt between the planets Mars and Jupiter. Like the planets, asteroids orbit the sun.

Asteroids come in many sizes. The largest is Ceres, which is about 600 miles across. (That's a little less than one-third the size of our moon.) Most asteroids, however, are less than a mile across. Astronomers estimate that there may be more than a million asteroids in the asteroid belt, with dozens of new ones being discovered every year. The largest known asteroids are listed below.

Sometimes asteroids fall out of their orbits due to a collision with another asteroid, a loss of orbital speed, or the gravitational pulls of the planets Mars and Jupiter. When that happens, these asteroids sometimes head on a collision course with Earth. These bodies, which have not yet entered Earth's atmosphere, are known as meteoroids. Meteoroids are rocky fragments that frequently range in size from several inches across to as small as a grain of rice. A few, however, may be quite large.

Name	Diameter
Ceres	600 miles
Vesta	360 miles
Pallas	330 miles
Hygeia	269 miles
Davida	240 miles

Or, these meteoroids may come close enough to Earth to be affected by its gravitational pull. As a result, they are pulled into Earth's atmosphere. Because they are traveling at high rates of speed, friction makes them and the air molecules around them heat up tremendously. As a result, most of these meteoroids burn up and disintegrate.

Fantastic Fact

Some scientists have estimated that at least 100 tons (that's 200,000 pounds) of meteoroids fall into Earth's atmosphere every day!

On clear nights, these meteoroids produce brilliant flashes of light. These flashes of light are called meteors. This is what we see when we say that there is a "shooting star." On most nights, you can see about three meteors each hour in the early evening and about six each hour between midnight and dawn. Remember that a meteor is *not* a solid object, but rather a flash of light.

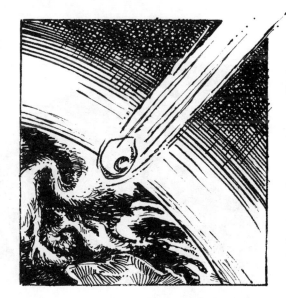

Often, in its annual path around the sun, Earth travels through a swarm of meteoroids. The glow from all these meteoroids, particularly at night, is known as a meteor shower. These showers are sometimes referred to as "cosmic fireworks" because they look like the brilliant sparklers and exploding lights of a Fourth of July fireworks show.

Across the Sky

Meteor showers are predictable events, occurring at regular times throughout the year. The chart below indicates some of the major annual meteor showers, the dates they normally occur, and the maximum number of meteors you can expect to see in an hour.

Shower	Dates	Max. number/hour
Quadranids	January 1–4	50
Lyrids	April 2–22	10
Eta Aquarids	May 3–5	10
Delta Aquarids	July 26–30	25
Perseids	August 9–15	50
Orionids	October 20–24	20
Leonids	November 14–20	10
Geminids	December 10–13	50

Seeing a meteor shower doesn't require a lot of preparation or expensive equipment. In fact, you will enjoy the "show" more if you don't use a telescope. Plan to go outside at night when there is very little light coming from the moon. (The darker the sky, the easier it is to see the meteors because many of them are very faint.) You'll be able to see more meteors after midnight than you will in the early hours of the evening.

You may find it easier to lie down on a blanket to look up at the sky. This helps keep your head still and you'll be able to focus on a larger part of the night sky. Look carefully, and you'll probably notice that the meteors tend to originate from the same area of the sky.

A meteor shower gets its name from the constellation nearest the part of the sky where it originates. It is important to keep in mind that the number of meteors in a meteor shower may vary from year to year. In some years there may be lots of meteors in a shower, while in other years there may be few.

Meteorites

Although most meteoroids burn up in Earth's atmosphere, a few survive the trip and strike Earth. A meteoroid that strikes Earth is known as a meteorite. Most of these are quite small; a few are very large. When a large meteorite strikes Earth, it does so with such force that it creates a hole in the ground, known as a crater. Often these meteorites strike Earth in unpopulated areas, but sometimes they fall in populated areas as well.

 ## Meteorite Search

You'll need:

large horseshoe magnet
string
plastic sealable sandwich bag
hand lens or microscope

What to do:

1. Visit a beach, lake front, or other sandy area.
2. Tie a piece of string to the middle of your horseshoe magnet and drag it through the sand.
3. Place the items that are attracted to the magnet into the sealable sandwich bag.
4. When you return home, use a hand lens or microscope to observe the tiny items you have collected.

What happens:

You'll notice that you have collected many different items with your magnet. Most of the items are quite small and may be dustlike in their appearance. It is estimated that up to 10 percent of those magnetic particles may be micrometeorites: very tiny meteorites. That's because tons and tons of meteorites rain down on the surface of Earth every day. Also, because one of the principal elements in meteorites is iron, they are attracted to the magnet.

Another place to search for micrometeorites is the dry lake beds and desert areas of the American Southwest. Iron-bearing rocks are less prevalent in those regions and the particles that are attracted to the magnet have a higher likelihood of being micrometeorites.

If you live in an area of the country that receives substantial amounts of snow in the winter, you may wish to gather a bucket of snow and allow it to melt. Dip your magnet into the melted snow and see if any potential celestial objects are attracted to it.

The most famous meteorite crater in the world is located in northern Arizona: the Barringer Meteorite Crater. It was formed when an enormous meteorite slammed into Earth many thousands of years ago. This crater is 4,150 feet across and 570 feet deep, which means the meteorite that created it was gigantic!

Fantastic Fact

The largest known meteorite, the Hoba West meteorite, fell in prehistoric times in what is now the country of Namibia in Africa. This chunk of rock weighs more than 60 tons and can still be seen today.

For years, scientists have debated several reasons for the extinction of the dinosaurs. One theory supports the idea that a giant asteroid slammed into Earth about 65 million years ago. Off the coast of Yucatan, Mexico, is a 110-mile-wide undersea crater. Some scientists believe that this crater was created by a 10-mile-wide asteroid that hit Earth with a violent impact. Dust and debris from that explosion would have blotted out the sun for many years. As a result, much of Earth's vegetation would have died off, interrupting the food chain on which dinosaurs depended. Consequently, the dinosaurs would have died off. Although there is still some debate about this theory, recent core samples taken from the crater and subjected to intensive laboratory testing indicate that the crater was, indeed, created at about the time the dinosaurs died out. What do you think?

Astronomers calculate that very large asteroids collide with Earth only about every hundred thousand years or so. Still, there is the chance that smaller asteroids, often several hundred yards in diameter, might hit Earth sooner than we would like to think. Although some recent movies would have us believe that this is practically a certainty when it is not, the reality is that it *could* occur. In fact, in 1994 an asteroid known as 1994 XL1 came within 65,000 miles of Earth. This is what scientists refer to as a "near miss."

Although 65,000 miles may seem like a long distance, keep in mind that it is only about one-fourth of the distance between Earth and the moon. Or, to look at it another way, it's the distance the average person drives his or her car in four years. Following are some other "near misses" in recent years.

Name of asteroid	Date closest to Earth	Distance from Earth
1994 XL1	December 9, 1994	65,000 miles
1993 KA2	May 20, 1993	93,000 miles
1991 BA	January 18, 1991	93,000 miles
4581 Asclepius	March 1, 1989	372,000 miles
Hermes	October 30, 1937	558,000 miles

Comets

Simply defined, comets are dirty snowballs. They are typically combinations of frozen water and gases, along with some dust and other solid particles, that orbit the sun. Many scientists believe that comets are frozen remainders of the cloud of material that formed the solar system billions of years ago.

Comets have four distinctive parts or elements:

- *Nucleus.* This is the center part of a comet. Made of a mixture of ice and dust, it is actually the smallest part of the comet. (The nucleus of Halley's Comet was measured in March 1986 by the spacecraft *Giotto*. It measured approximately 10 miles long by 5 miles wide.)
- *Coma.* This is an envelope of gas that completely surrounds the nucleus of the comet. It's formed when the comet approaches the sun and the frozen surface starts to evaporate, forming a great head of gas. The coma also reflects the light of the sun and is often much larger than Earth.
- *Gas tail.* One of the most distinctive parts of a comet, the gas tail is straight and narrow, forced back by electrically charged particles in the solar wind. The tail is longest at the comet's closest approach to the sun. Some comet tails may be tens of millions of miles long.
- *Dust tail.* This part of the comet's tail follows the curve of the comet's path as the comet approaches the sun. It consists primarily of dust and gases pushed away by the force of the solar wind.

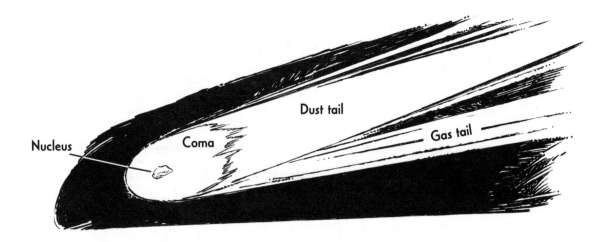

Nucleus

Coma

Dust tail

Gas tail

Fantastic Fact

Comets travel tail-first when they're receding away from the sun. That's because the tail is always pushed away from the sun by the solar wind.

Comets often travel in long orbits around the sun and out into the farthest reaches of the solar system. Occasionally, some comets get too close to the sun and are destroyed. Others may pass by the sun once, then speed out of the solar system, never to be seen again. And a few go around the sun every few years or so. Scientists have roughly classified comets into three broad categories:

- *Nonperiodic comets.* These are comets whose orbit around the sun can take millions of years. Delavan's Comet, seen in 1914, will not reappear in our solar system for an estimated 24 million years.
- *Long-period comets.* These comets may take anywhere from 200 years to a few thousand years to orbit the sun. The Hale-Bopp Comet flew by about 120 million miles from Earth in March 1997. People were able to see it without the aid of a telescope. However, it won't return again for more than 2,300 years.
- *Short-period comets.* These comets are those that are trapped by the gravity of planets, particularly Jupiter, the largest planet. They tend to orbit the sun relatively quickly. Comet Encke, for example, takes only 3.3 years to orbit the sun.

Comet Websites

Following are some terrific websites at which you can learn lots of incredible things about comets.

- http://medicine.wustl.edu/~kronkg/index.html
- http://www.hawastsoc.org/solar/eng/comet.htm
- http://www.pbs.org/wgbh/nova/spacewatch
- http://www.jpl.nasa.gov/comet
- http://seds.lpl.arizona.edu/sl9/sl9.html
- http://exn.ca/space/stardust/index.cfm

A Most Famous Comet

For a long time people thought that comets were random astronomical events. That is, comets came and went with no recognizable pattern to their appearance or disappearance.

In 1682 the English astronomer Edmond Halley sighted a comet through his telescope. He was the first to calculate that three separate comets that had been seen over the years were, in fact, the same comet, the one Halley "discovered." Halley then correctly calculated that the comet he saw was the same one seen in the year 1531 and again in 1607. That meant that the comet had an orbit around the sun of once every 76 years.

Halley predicted that the comet would reappear again in the year 1758, which it did! As a result of his work, the comet was named after him and is now known as Halley's Comet, perhaps the most famous comet in modern times. It always appears on a regular and predictable basis: every 76 years.

Fantastic Fact

Mark Twain, the famous writer, was born on November 30, 1835. Halley's Comet, a comet that passes by Earth once every 76 years, was seen that year. During the later years of his life, Mark Twain used to say: "I came in with Halley's Comet in 1835 . . . and I expect to go out with it." Amazingly, he was right. Halley's Comet appeared in 1910. Mark Twain died on April 21, 1910. What an incredible coincidence!

Following are the dates of some of the past and future flybys of Halley's Comet.

1531	1910
1607	1986
1682	2062
1758	2138
1835	2214

English astronomer Edmond Halley

Note: Interestingly, the comet that now bears Edmond Halley's name has passed by Earth 30 times since 217 B.C.

In ancient times, many people around the world associated comets with evil happenings and bad times. It was thought that when a comet appeared, terrible things would happen: There would be great natural disasters, plagues, or the death of large numbers of people. Today we know that comets are natural phenomena that often appear in the skies on regular and predictable timetables. And they are just another of the wonderful discoveries and events in our universe.

7

Visiting a Planetarium

A comet streaks across the southern sky in a long and graceful arc. A space probe swings around Earth to gain speed from the gravitational pull of this third planet from the sun. A small shoebox-sized, motorized vehicle makes a bumpy landing on a distant planet and begins creeping across its surface. Several humans work together to open the arms of a solar panel on a high-orbiting space station. A never-before-seen star dies.

Where is the place where we're making these observations? It's not part of some science fiction novel or comic book story. In fact, it may be just around the corner, down the highway, or a short vacation away. It's a planetarium: a place where dozens of activities and millions of celestial bodies await your discovery. Planetariums (or observatories) can be found in nearly every state, from those along the Pacific and Atlantic Oceans to those far inland on the plains and mountains of middle America.

What Is a Planetarium?

A planetarium is a domed theater (usually with a blackened ceiling) in which a scientifically accurate simulation of the night sky is created. This is done through the use of a "star projector," a machine that projects dots of light on the ceiling. The projector is also able to project the movement and positions of the sun, moon, and planets throughout the solar system. Star projectors can also show the correct sky for any date of the year and for any location on the surface of the Earth.

Most planetarium programs last 30 to 40 minutes. They are usually about astronomy or some astronomy-related topic. Typically, they project stars, constellations, and other celestial bodies on a curved surface. Often, a narrator tells stories about the universe, including legends and current research and discoveries. Special effects may be used to depict sunrises, sunsets, rainbows, constellation figures, and animated objects such as meteors and comets. Frequently, a planetarium is able to project specialized objects such as rotating planets, multiple star systems, galaxies, or black holes.

Most planetariums design their own special effects. These effects may be produced through the use of video, computer graphics, and laser technology. These effects allow visitors to see celestial objects in sharp detail and frequently in a time-compressed mode. (For example, although it takes Earth 365 days to orbit the sun, a planetarium can depict that orbit in a matter of minutes.)

A visit to a planetarium can be a magical and marvelous experience. Everything you have been learning in this book will come to life as you tour the numerous exhibits and observe the incredible array of information available. Hundreds of celestial bodies, scores of constellations, millions of stars, loads of scientific equipment, and dozens of hands-on experiences with every facet of the universe are some of the highlights that planetariums throughout the country have to offer. Eye-popping exhibits and mind-bending displays provide you with rare opportunities to come face-to-face with the incredible universe in which you live. Planetariums are places of wonder, places where you can stroll, meander, skip, or dart from planets to stars and back again, all at your own pace. Even if you never have an opportunity to journey into space or see a solar eclipse, you will definitely want to plan a visit to a planetarium.

Following is a list of selected planetariums around the United States. If you're not able to visit these planetariums in person, you may want to contact them by mail to request their literature or informational brochures. Several planetariums have their own websites, which can provide you with numerous opportunities to take an armchair journey through their exhibits and displays.

Pacific States

W. M. Keck Observatory
65-1120 Mamalahoa Hwy.
Kamuela, HI 96743
(808) 885-7887
Website: http://www2.keck.hawaii.edu:
   3636

Morrison Planetarium
California Academy of Sciences
Golden Gate Park
San Francisco, CA 94118
(415) 221-5100
Website: http://www.calacademy.org/
   planetarium

Chabot Observatory and Science Center
4917 Mountain Blvd.
Oakland, CA 94619
(510) 530-3480

Stellarium
4560 Petaluma Hill Rd.
Santa Rosa, CA 95404
(707) 586-0660
Website: http://www.stellarium.com

Discovery Museum Learning Center
3615 Auburn Ave.
Sacramento, CA 95628
(916) 277-6181

Pacific States, cont.

Griffith Observatory
2800 E. Observatory Rd.
Los Angeles, CA 90027
(323) 664-1181
Website: http://griffithobs.org

Mount Wilson Observatory
P.O. Box 60947
Pasadena, CA 91116
(626) 793-3100
Website:http://www.mtwilson.edu/
    index.html

Fleet Space Theater and Science Center
1875 El Prado
San Diego, CA 92101
(619) 238-1233
Website: http://www.rhfleet.org

Oregon Museum of Science and Industry
1945 South East Water Ave.
Portland, OR 97214
(503) 797-4000
Website: http://www.omsi.edu

Pacific Science Center
200 2nd Ave., North
Seattle, WA 98109
(206) 443-2001
Website: http://www.pacsci.org

Southwestern States

Arizona Science Center
600 E. Washington
Phoenix, AZ 85004
(602) 716-2000
Website: http://www.azcentral.org

Flandrau Science Center and Planetarium
University of Arizona
Cherry Ave. & University Blvd.
Tucson, AZ 85721
(520) 621-4515
Website: http://www.flandrau.org

Kitt Peak Museum
Tucson, AZ 85735
(602) 322-3426

Lowell Observatory
1400 W. Mars Hill Rd.
Flagstaff, AZ 86001
(520) 774-3358
Website: http://www.lowell.edu

The Space Center
Tombaugh Planetarium
P.O. Box 533
Alamagordo, NM 88311
(505) 437-2840
Website: http://www.zianet.com/space

Mountain States

Denver Museum of Natural History
Gates Planetarium
2001 Colorado Blvd.
Denver, CO 80205
(303) 370-6357
Website: http://www.dmnh.org

The Discovery Center of Idaho
131 Myrtle St.
Boise, ID 83702
(208) 343-9895
Website: http://scidaho.org

Hansen Planetarium
15 S. State St.
Salt Lake City, UT 84111
(801) 538-2104
Website: http://www.utah.edu/planetarium

Midwestern States

Science Center of Iowa
4500 Grand Ave.
Greenwood-Ashworth Park
Des Moines, IA 50312
(515) 274-6868
Website: http://www.sciowa.org

Kansas Cosmosphere and Space Center
1100 N. Plum
Hutchinson, KS 67501
(316) 662-2305
Website: http://www.cosmo.org/
    education.htm

Lake Afton Public Observatory
MacArthur Rd. at 247th St., W
Wichita, KS 67260
(316) 794-8995
Website: http://www.twsu.edu/
    ~obswww

Kansas City Museum
3218 Gladstone Blvd.
Kansas City, MO 64123
(816) 483-8300
Website: http://www.kcmuseum.com/
    index02.html

St. Louis Science Center
5050 Oakland Ave.
St. Louis, MO 63110
(314) 289-4444
Website: http://www.slsc.org

Hastings Museum
1330 N. Burlington
Hastings, NE 68902
(402) 461-4629

University of Nebraska State Museum
Morrill Hall
14th & U Sts.
Lincoln, NE 68588
(402) 472-2637

South Dakota Discovery Center and
    Aquarium
805 W. Sioux Ave.
Pierre, SD 57501
(605) 224-8295

Southern States

Lafayette Natural History Museum,
　Planetarium, and Nature Station
637 Girard Park Dr.
Lafayette, LA 70503
(318) 268-5544
Website: http://www.lnhm.org/planet/
　chrtjune.htm

Omniplex Science Museum
2100 N.E. 52nd St.
Oklahoma City, OK 73111
(405) 424-5545

Harrington Discovery Center
1200 Streit Dr.
Amarillo, TX 79106
(806) 355-9547

Fort Worth Museum of Science and
　History
1501 Montgomery St.
Fort Worth, TX 76107
(817) 732-1631

Houston Museum of Natural Science
One Hermann Circle Dr.
Hermann Park, TX 77030
(713) 639-4600

The Science Place
1318 2nd Ave.
Dallas, TX 75315
(214) 428-7200
Website: http://www.scienceplace.org/
　discover.htm

Space Center Houston
1601 NASA Road One
Houston, TX 77058
(713) 244-2105
Website: http://www.spacecenter.org

Great Lakes States

The Adler Planetarium
1300 S. Lake Shore Dr.
Chicago, IL 60605
(312) 922-STAR

The Children's Museum of Indianapolis
3000 N. Meridian St.
Indianapolis, IN 46208
(317) 924-5431

Abrams Planetarium
Michigan State University
East Lansing, MI 48824
(517) 355-4672
Website: http://www.pa.msu.edu/
　abrams/

Cranbrook Institute of Science
1221 N. Woodward Ave.
Bloomfield Hills, MI 48303
(313) 645-3230
Website: http://www.cranbrook.edu/
　institute/programs.html

Kalamazoo Public Museum
315 S. Rose St.
Kalamazoo, MI 49007
(616) 345-7092

Kingman Museum of Natural History
W. Michigan Ave. at 20th St.
Battle Creek, MI 49017
(616) 965-5117

Great Lakes States, cont.

Michigan Space Center
Jackson Community College
2111 Emmons Rd.
Jackson, MI 49201
(517) 787-4425

Cincinnati Museum of Natural History
and Planetarium
Museum Center at Union Terminal
1301 Western Ave.
Cincinnati, OH 45203
(513) 287-7020

Cleveland Museum of Natural History
One Wade Oval Dr.
University Circle
Cleveland, OH 44106
(216) 231-4600
Website: http://www.cmnh.org/
cmnh.html

COSI/Columbus—Ohio's Center of
Science and Industry
280 E. Broad St.
Columbus, OH 43215
(614) 228-2674

The Dayton Museum of Natural History
2600 DeWeese Pkwy.
Dayton, OH 45414
(513) 275-9156

Southeastern States

U.S. Space and Rocket Center
One Tranquillity Base
Huntsville, AL 35805
(800) 63-SPACE

Discovery Science Center of Central
Florida
50 S. Magnolia Ave.
Ocala, FL 34474
(904) 620-2555

Miami Museum of Science and Space
Transit Planetarium
3280 S. Miami Ave.
Miami, FL 33129
(305) 854-4247

Museum of Arts and Sciences
1040 Museum Blvd.
Daytona Beach, FL 32114
(904) 255-0285

Museum of Science and History
1025 Museum Cir.
Jacksonville, FL 32207
(904) 396-7062

Museum of Science and Industry
4801 E. Fowler Ave.
Tampa, FL 33617
(813) 987-6324

Orlando Science Center
810 E. Rollins St.
Loch Haven Park
Orlando, FL 32803
(407) 896-7151
Website: http://www.ocs.org

The Science Center of Pinellas County
7701 22nd Ave., N
St. Petersburg, FL 33710
(813) 384-0027

Southeastern States

Fernbank Science Center
156 Heaton Park Dr., NE
Atlanta, GA 30307
(404) 378-4311
Website: http://fsc.fernbank.edu

Savannah Science Museum
4405 Paulsen St.
Savannah, GA 31405
(912) 355-6705

Discovery Place
301 N. Tryon St.
Charlotte, NC 28202
(704) 372-6261

Morehead Planetarium
E. Franklin St.
Chapel Hill, NC 27599
(919) 549-6863
Website: http://ils.unc.edu/toch/
    planet.html

The Natural Science Center of Greensboro
4301 Lawndale Dr.
Greensboro, NC 27408
(919) 288-3769
Website: http://www.greensboro.com/
    sciencecenter

Schiele Museum of Natural History and
    Planetarium
1500 E. Garrison Blvd.
Gastonia, NC 28054
(704) 866-6900

SciWorks
400 Hanes Mill Rd.
Winston-Salem, NC 27105
(919) 767-6730
Website: http://www.sciworks.org/
    planetarium.htm

Museum of York County
4621 Mount Gallant Rd.
Rock Hill, SC 29732
(803) 329-2121

Roper Mountain Science Center
504 Roper Mountain Rd.
Greenville, SC 29615
(803) 281-1188

Memphis Pink Palace Museum and
    Planetarium
3050 Central Ave.
Memphis, TN 38111
(901) 320-6320
Website: http://www.memphisguide.com/
    ppalace.html

Science Museum of Virginia
2500 W. Broad St.
Richmond, VA 23220
(804) 367-1013
Website: http://world.smv.mus.va.us/
    wunihome.html

Science Museum of Western Virginia
One Market Square
Roanoke, VA 24011
(703) 342-5710
Website: http://www.cits.org/scimuse.html

Virginia Air and Space Center
600 Settlers Landing Rd.
Hampton, VA 23669
(804) 727-0800

Middle Atlantic States

National Air and Space Museum
Smithsonian Institution
6th & Independence, SW
Washington, DC 20560
(202) 357-1300

Owens Science Center
9601 Greenbelt Rd.
Lanham, MD 20706
(301) 918-8750
Website: http://www.gsfc.nasa.gov/
hbowens/planetarium.html

Maryland Science Center
601 Light St.
Baltimore, MD 21230
(410) 685-5225
Website: http://www.mdsci.org

The Newark Museum
Dreyfuss Planetarium
49 Washington St.
Newark, NJ 07101
(201) 596-6550
Website: http://rutgers.newark.rutgers.edu/
dreyfuss/

American Museum—Hayden Planetarium
81st St. & Central Park, W
New York, NY 10024
(212) 769-5920
Website: http://www.amnh.org/rose/

Museum of Science and Technology
500 S. Franklin St.
Syracuse, NY 13202
(315) 425-9068

Roberson Museum and Science Center
30 Front St.
Binghamton, NY 13905
(607) 772-0660
Website: http://www.roberson.org

Rochester Museum and Science Center
657 East Ave.
Rochester, NY 14607
(716) 271-4320
Website: http://www.rmsc.org

Schenectady Museum and Planetarium
Nott Terrace Heights
Schenectady, NY 12308
(518) 382-7890
Website: http://www.cyhaus.com/
museum

Carnegie Science Center
One Allegheny Ave.
Pittsburgh, PA 15212
(412) 237-3300
Website: http://csc.clpgh.org

Franklin Institute Science Museum
20th St. & Benjamin Franklin Pkwy.
Philadelphia, PA 19103
(215) 448-1200
Website: http://www.fi.edu

Northeastern States

The Discovery Museum
duPont Planetarium
4450 Park Ave.
Bridgeport, CT 06604
(203) 372-3521
Website: http://www.discoverymuseum.
  org/planetarium.htm

Science Center of Connecticut
950 Trout Brook Dr.
West Hartford, CT 06119
(203) 231-2824
Website: http://www.sciencecenterct.org

Museum of Science
Science Park
Boston, MA 02114
(617) 589-0100
Website: http://go.boston.com/mos

New England Science Center
222 Harrington Way
Worcester, MA 01604
(508) 791-9211
Website: http://nesc.org/index.html

Museum of Natural History
Roger Williams Park
Providence, RI 02907
(401) 785-9450

The Discovery Museum
51 Park St.
Essex Junction, VT 05452
(802) 878-8687

Fairbanks Museum and Planetarium
Main & Prospect Sts.
Johnsbury, VT 05819
(802) 748-2372
Website: http://www.pbpub.com/
  museum.htm

Special Note:

**The Online Planetarium Show
  website:**
(http://library.advanced.org/3461/
  toc_f.htm)
This site provides an opportunity to
"visit" a working planetarium from the
comfort of your own home. It is truly
incredible and offers tons of fascinating
and amazing information, photos, real-
time videos, and more about the universe. Don't miss it!

The Dome of the Sky website:
(http://einstein.stcloudstate.edu/dome/
  default2.html)
Another super online planetarium. Here
you can search all 88 constellations *and* their
history and legends, as well as find photos
and illustrations of each one. Super!

    Following are addresses of two other planetariums around the world that
you may be able to visit someday.

Northern Lights Planetarium
Nordlysplanetariet Tromsø as
  Universitetsområdet, Breivika
9037 TROMSØ
Norway
Website: http://www.uit.no/planetarium

Royal Observatory Greenwich and
  Planetarium
National Maritime Museum
Greenwich
London, England SE10 9NF
Website: http://www.nmm.ac.uk/tm/
  oro.html

Planetariums and observatories are noted for their spectacular shows of stars, constellations, and other celestial bodies. You can create your own miniature observatory right at home with the following activity.

Constellations in a Can

You'll need:

12 black plastic 35mm film canisters (you can obtain these free by visiting a local photo or camera store)
scissors
push pin
cellophane tape
photocopy of page 84 (12 constellations)
paper

Note: You should get an adult to assist you in preparing the materials for this activity.

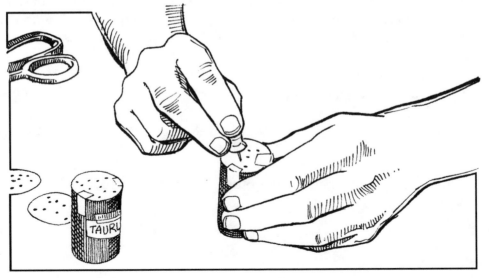

What to do:

1. From the photocopy, cut out each of the 12 constellations in a circle shape with a pair of scissors.

   Note: All the constellations have been printed the reverse of the way they appear in the sky. When you create the constellations for this activity, however, they will appear in their correct placement.

2. Tape one circle on the outside bottom of a film canister. Use a push pin (invite an adult to help you) to poke holes through the bottom of the canister, one for each dot (star) in the constellation. Do not hold the canister over your lap or up to your face when doing this. Be sure that you push the pin though the canister over a solid surface in a downward motion. After poking all the holes in a canister, remove the circle from the bottom.

3. Repeat for each of the other canisters and constellations.

4. Write the name of each constellation on a small piece of paper and tape the paper to the side of the appropriate canister.

5. Hold each canister up to a light source (overhead light, lamp) and observe the constellation through the open end (each constellation will now appear as it does in the night sky).

6. *Optional:* Create additional canisters for additional constellations, and invite family members to view the constellations in your homemade planetarium. How many of the 88 constellations can you create?

What happens:

You have created miniature replicas of some major constellations. It is important to keep in mind that your "constellation canisters" are two-dimensional replicas of three-dimensional objects in space. In other words, the stars in a constellation are not all the same distance away from Earth (as they are in your homemade planetarium). The stars in a constellation are located at various light-years' distance from Earth and from one another. Although you are able to see a single constellation in a canister, it is also important to remember that there are millions of other stars in the night sky in and around the "stars" you see.

After you have had an opportunity to practice with your own homemade planetarium, go out on a warm night and try to locate your identified constellations in the sky above.

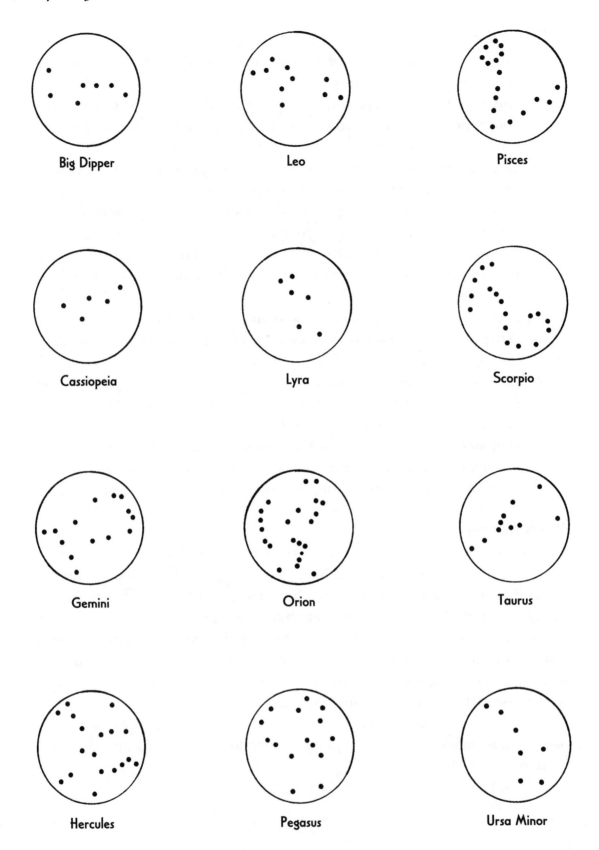

Big Dipper

Leo

Pisces

Cassiopeia

Lyra

Scorpio

Gemini

Orion

Taurus

Hercules

Pegasus

Ursa Minor

Planetarium Programs

Following is a selection of planetarium programs from a few planetariums around the United States. This collection is designed to illustrate the incredible discoveries awaiting you at your local planetarium as well as the rich variety of shows for every age group and every astronomer, young or old. The complete addresses for these planetariums are in the list earlier in the chapter.

duPont Planetarium
The Discovery Museum • Bridgeport, Conn.

"EXPLORING THE UNIVERSE"

How about a trip to the planets? On the darkened dome overhead, visitors watch as the images start at the planetarium and move out through the known universe. They visit Mars and the other planets in our solar system, the sun, and the moon, learning the latest information from recent explorations. They also watch as the constellations move into their nighttime positions. The show takes advantage of the facility's new computer system and specialized audio and visual effects.

"COMETS: FROM FIRE TO ICE"

Ride on a manned landing craft about to dock on Halley's Comet. Multiple images appear and dissolve overhead: meteor showers, the vast cloud of giant icy snowballs from which comets form, and historical images of comets. Finding out how they evolved, what they are made of, and the history of comets is all part of this incredible show.

Tombaugh Planetarium
The Space Center • Alamagordo, N.Mex.

PLANETARIUM OVERVIEW

Built in 1980, this 40-foot-diameter tiled dome theater and planetarium houses a Spitz 512 projector, which can project 2,400 stars, the five visible planets, and the sun and the moon. An IMAX dome projection system projects the movie and sound around each and every observer. Each program lasts for about 30 minutes.

MUSEUM GALLERY 4A

This gallery contains exhibits on the first manned space flights, both U.S. and Soviet. Mercury, Gemini, and Apollo capsule models are featured. Also featured is the Lunar Rover, which enabled the astronauts to travel greater distances on the moon and carry back hundreds of pounds of lunar rock samples. Each artifact is highlighted by a spotlight as narration describes the object and its relevance to the space program.

Morrison Planetarium
California Academy of Sciences • San Francisco, Calif.

"STARDUST"

Take a closer look at some of the minor bodies orbiting our star—from particles of dust to rocky asteroids and icy comets—and find out how some of the solar system's smallest members can have the greatest impact on us.

"THE SKY TONIGHT"

This is a relaxing tour of the current night sky, focusing on constellations, visible planets, the phase of the moon, and upcoming celestial events visible with the naked eye. This is a free-form show where the sky is the main effect.

 Star Party

You'll need:

> sheet of clear acetate (sometimes called a transparency master), available at any office supply store
> four permanent felt-tip markers (black, red, green, blue)
> cellophane tape
> masking tape

What to do:

1. Locate a windowpane in your house, one that has a clear line of sight to the sky (no trees or other objects in the way).
2. Stand close to the window and tape the sheet of acetate to the glass slightly above the level of your eyes.
3. Stand back about one foot and place a strip of masking tape on the floor.
4. On a clear (preferably moonless) night when several stars are clearly visible, stand with your toes on the strip of masking tape. Look out the window through the sheet of acetate taped to the glass.
5. Locate several prominent stars (see Chapter 5, "Stars and Constellations"). Use the black marker to mark the location of six or seven of those stars directly on the sheet of acetate (make solid dots). Write the time and date at the bottom of the sheet.

6. One week later, stand at the same place (with your toes on the masking tape) at the same time. Note the position of the same stars and mark each one on the acetate with the red marker. Write the date on the sheet.

7. One week later, stand at the same place (with your toes on the masking tape) at the same time. Note the position of the same stars and mark each one on the acetate with the green marker. Write the date on the sheet.

8. One week later, stand at the same place (with your toes on the masking tape) at the same time. Note the position of the same stars and mark each one on the acetate with the blue marker. Write the date on the sheet.

What happens:

You will note that the marks you placed on the sheet of acetate are different for each week, even though you were using the same stars each time. You may think this means that the stars have moved in the sky. Actually, the stars have remained in stationary positions in the sky: It is Earth that has moved. As Earth orbits the sun, it changes its position in the sky relative to the stars. So, each week Earth moves a little farther in its path, while the stars remain "fixed" in their positions. (In one week, Earth travels more than 9 million miles in its orbit; in four weeks it will have traveled more than 36 million miles.

Planetariums and the shows they offer are as varied and as different as the celestial bodies that exist in our universe. There are enormous planetariums spreading over several acres of land, and there are smaller observatories of only a few rooms and a couple of exhibits. Some planetariums have elaborate and complex shows; others are simpler and more modest. Actually, no two planetariums are the same, but they all share one vital function: to provide a glimpse into the mysteries and marvels of the universe and its incredible wonders. A primary goal of any planetarium is to help the public better understand our world and its place in the vastness of space.

Planetariums provide us with opportunities to examine, explore, and learn about the discoveries made by working scientists around the world. They also allow us to make our own discoveries about the universe. Planetariums are magical places that invite questions, provide answers, and stimulate numerous investigations about Earth and the sky. Plan to visit one in person or via the Internet to learn more about the solar system and the universe.

8

Space Exploration

E ver since human beings first looked up at the skies, we have wondered about the moon, the sun, the planets, and other celestial bodies. What's out there? What can we learn? Why do certain events happen the way they do? Is there intelligent life somewhere else in the universe? These are all questions that have been asked for many thousands of years. For some of them we have learned the answers. For others the answers are still waiting to be discovered.

Only in the last 40 years or so have we had the technical knowledge to seek answers to those queries and a thousand other questions. We have sent hundreds of satellites, probes, spacecraft, rockets, space shuttles, and other objects into space to learn more about our fabulous universe. Some of these missions have been manned, providing human beings with the first opportunities to physically leave Earth's gravity and visit places in space that people previously could only dream about.

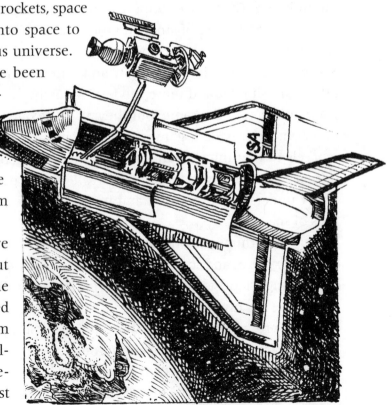

With our technology we have been able to learn about the moon, the sun, and all the planets. We have photographed stars billions of miles away from Earth, set foot on Earth's satellite, and sent specialized vehicles to roam across the vast landscapes of distant worlds. We have been able to build enormous space stations that orbit Earth and provide us with opportunities to learn more about our world and its place in the solar system.

And who knows what kinds of adventures, discoveries, or missions await us in the future? Currently there is talk of a manned colony on the surface of the moon as well as one on the planet Mars. There are plans to send spacecraft to far-distant objects in our solar system to gather materials for research and examination here on Earth. And there are discussions about deep space missions that may last for several years, decades, or even longer. It's sometimes amazing to think that what once was science fiction is now becoming science fact. The only limits are the limits of our own imaginations. In space exploration, everything is possible, nothing is impossible.

The Space Age

October 4, 1957, is considered by many to be the start of the space age. It was on this date that the Soviet Union launched a tiny artificial satellite known as *Sputnik 1*. This first space object launched by humans was nothing more than a small metal sphere with four antennae and a compact radio transmitter. As small as it was, it was perhaps the single most important object ever placed into space, because it was the first.

Four years after the launch of *Sputnik*, Yuri Gagarin became the first human being to be launched into space. Since then there have been countless expeditions, discoveries, and missions beyond our atmosphere. In the intervening years we have explored and discovered much about space. We have answered some old questions and posed many new ones. Complex spacecraft and the related advanced technology have enlightened and enriched us.

In the early years of the twenty-first century, we are planning to send space shuttles regularly into space to serve as orbiting laboratories. Nations are banding together to sponsor joint missions, reaping the benefits of space exploration in a cooperative spirit of camaraderie and friendship. Space is often called "The Final Frontier," and it may be the most exciting one.

The chart on page 90 lists selected events and milestones in space exploration. It is not meant to be all-inclusive, but rather an indicator of the giant leaps and great strides we have made in space exploration. Further information and more details can be obtained from the many websites listed in Chapter 9.

Date	Spacecraft	Description
October 4, 1957	*Sputnik 1*	Was first satellite launched into space
September 15, 1959	*Luna 2*	Was first rocket to reach the moon
April 12, 1961	*Vostok 1*	Took first human into space
December 14, 1962	*Mariner 2*	Flew past Venus
July 31, 1964	*Ranger 7*	Took close-up photos of the moon's surface
July 14, 1965	*Mariner 4*	Did flyby of Mars
March 1, 1966	*Venera 3*	Landed on Venus
July 20, 1969	*Apollo 11*	Was first manned landing on the moon
April 1971	*Salyut 1*	Was first space station
December 3, 1973	*Pioneer 10*	Flew past Jupiter
July 20, 1976	*Viking 1*	Landed on Mars
September 1, 1979	*Pioneer 11*	Flew past Saturn
April 1981	*Columbia*	Was first space shuttle
January 24, 1986	*Voyager 2*	Flew past Uranus
March 6, 1986	*Vega 1*	Took photos of Halley's Comet
August 25, 1989	*Voyager 2*	Flew past Neptune
July 4, 1997	*Pathfinder*	Was first rover on surface of Mars

Space travel is made possible through the use of rocket power. As you might imagine, it takes tremendous power to escape from Earth's gravity. It's the force of gravity that holds objects down to Earth's surface by pulling them toward its center.

For an object to go into Earth's orbit, it needs to be traveling at a speed of about 18,000 miles per hour. For a spacecraft to completely escape from Earth's gravity, it needs to be traveling at a speed of approximately 25,000 miles per hour.

Rockets must be very powerful to escape the gravitational pull of Earth. Typically, a rocket consists of several stages, each with its own fuel source. This fuel is usually a combination of liquid hydrogen and liquid oxygen, which mix together in a chamber to create a continuous explosion. This exerts tremendous pressure on the walls of the chamber. At the end of the mixing chamber is an escape nozzle through which the exhaust gases are allowed to escape. This creates an enormous force at the front of the chamber (the increased internal pressure), which pushes the rocket forward. Following is an activity that demonstrates this action.

Rocket Power

You'll need:

9-inch round balloon
18-inch long balloon
5-ounce paper cup
scissors

What to do:

1. Use the scissors to cut the bottom from the paper cup.
2. Inflate the long balloon partially and pull the neck of the balloon through the bottom and out the top of the cup (see the illustration).
3. Fold the neck of the long balloon over the edge of the cup to prevent any air from escaping.
4. Place the round balloon (top first) into the open top of the cup and inflate it.
5. Let go of the neck of the round balloon.

What happens:

This simple device illustrates how a three-stage rocket works. Each stage of a rocket has its own fuel supply, which is used to lift and move it out of Earth's gravity and out into space. When a stage of the rocket uses up its fuel, it drops away, making the rocket lighter. In this example, the round balloon represents the first stage of your "rocket." After it has used up its "fuel" (the air inside it), it drops away. The "rocket" is now slightly lighter. The third stage (the long balloon) can now use its "fuel" (the air inside it) to propel the rocket even farther. (**Note:** The second stage of this "rocket," the paper cup, does not have a "fuel" supply.) In reality the rocket would have sufficient speed to go into orbit or completely leave Earth's atmosphere for a journey into space.

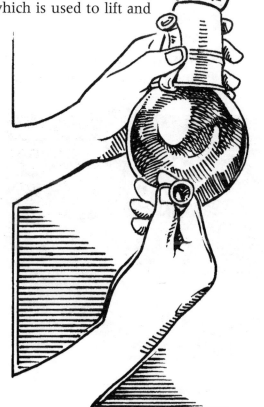

Satellites

Today there are more than 200 artificial satellites circling Earth. Satellites are designed to do a variety of jobs. A few are launched for scientific research. Others permit international phone calls to be made in microseconds. Some are designed to beam TV signals all around the globe. Special communications satellites can combine thousands of phone calls and TV signals and deliver them simultaneously to all parts of the world. Other satellites can forecast weather patterns, assist in the navigation of airplanes and ships at sea, or locate valuable deposits of oil and minerals for mining. Following are some of the different types of satellites.

- *Comsats.* These are communication satellites that relay phone and TV signals to every spot on the globe. One series of Comsat satellites, COMSTAR, can relay more than 18,000 phone calls at the same time all over the United States.
- *Landsats.* These satellites provide information on crops, natural disasters, deforestation areas, and changing coastlines.
- *Seasats.* These satellites gather information about the world's oceans, including changes in the sea bed, ocean currents, oceanic storms, and schools of fish.
- *Navistar.* This network of U.S. satellites is used by ships and airplanes to plot their location to within 75 yards of any place on Earth.
- *Sage.* This type of satellite is used to measure and monitor Earth's ozone layer.
- *Iras.* This type of satellite is used to detect infrared radiation from galaxies hundreds of millions of light-years away.

Satellites are categorized by the type of job they have to do. The four basic categories are listed in the following chart.

U.S. satellite Nimbus–7

Orbit	Description
Low Earth	Like its name, this type of satellite orbits Earth in a very low trajectory. The space shuttle and space stations typically use this type of orbit.
Polar	These types of satellites circle the globe in a north-south direction, passing over the two poles. These are the preferred satellites for spy missions and global surveys because they can cover the entire Earth as it turns.
Geostationary	Geostationary satellites continuously hover over one spot on Earth's surface. They are placed exactly 22,246 miles above the equator. Weather and communications satellites use geostationary orbits.
Eccentric	Eccentric satellites have an irregular orbit that moves them primarily over large land masses. This type of satellite is preferred by the Soviets.

The following activity will help you understand the movement of artificial satellites around our planet.

 # Circular Journey

You'll need:

    large plastic garbage bag
    scissors
    round garbage can
    piece of string
    baseball
    marble
    cardboard tube (from a roll of paper towels)

What to do:

1. Cut the plastic garbage bag open along the two sides so that it forms a large sheet.

2. With the help of a family member, stretch the single layer of plastic over the top of a round garbage can. Tie a piece of string around the top edge of the can to hold the plastic securely on the can. Make sure it is tightly stretched across the top.

3. Put a baseball in the middle of the plastic so that it forms a slight depression in the sheet. (The baseball simulates a planet.)

4. Hold the cardboard tube at about a 45-degree angle along one edge of the plastic film.

5. Roll the marble down the tube so that it will roll around the outside edge of the plastic sheet. (You may have to practice several times to achieve the desired effect.)

6. The marble will roll in concentric circles around the outer edge of the plastic sheet, moving closer and closer to the baseball in the middle.

What happens:

Satellites orbit (or revolve) around other objects in space (usually planets). In this activity, the baseball simulates a large planet. The marble simulates a satellite. When you roll the marble (satellite) onto the plastic sheet, it begins to move around and around the planet (baseball) in concentric circles. However, unlike satellites in space, the pull of gravity on the marble is greater than its speed. Consequently, the circles it makes become smaller and smaller until it falls into the baseball. In space, an object stays in orbit when its speed is sufficient to balance the gravitational pull of the object around which it is orbiting. If gravity is stronger than the orbital speed, the object is pulled inward (as in our demonstration). If the speed of the satellite is much greater than the force of gravity of the larger object (planet), the satellite speeds out into space. When scientists launch satellites to orbit Earth, they must be very precise in their calculations so that there is a "match" between the gravitational pull of Earth and the velocity of the artificial satellite.

Space Shuttles

One of the most exciting developments in recent years has been the creation and launch of the space shuttle. This flying machine was designed primarily to reduce some of the costs associated with space travel as well as to reduce the amount of material destroyed in a typical rocket launch.

At the end of the twentieth century there were four space shuttles in existence: *Atlantis, Columbia, Discovery,* and *Endeavor.* Each is pushed into space by two solid-fuel booster rockets that produce the power produced by 140 jumbo jets. Special design allows each shuttle craft to be used many times, taking heavy objects, such as satellites and space probes, around Earth, and carrying crews for scientific research into space.

Shuttle Facts

- Crew: 8 people
- Length: 122 feet
- Height: 56 feet
- Wingspan: 78 feet
- Weight: 173,000 pounds (approximately)
- Thrust: 393,800 pounds per engine at sea level
- Cargo bay: 60 feet long, 15 feet in diameter
- Approximate cost: $2.1 billion
- Orbital speed: 17,500 mph
- Orbit: 190 to 350 miles above sea level
- Payload: 29,500 kg
- Parts: three (an orbiter, an external tank, two solid rocket boosters)
- Mission duration: typically one week

If you would like to fly a space shuttle (via your computer), check out the following website: http://www.spacecamp.com.

Space Stations

Interestingly, space stations are also referred to as "satellites" because they are artificial objects put into space by humans to orbit Earth. Space stations are designed so that humans can live and work in space for long periods of time. Much of the work on a space station is scientific in nature, such as studying the effects of prolonged weightlessness on the human body, assembling scientific apparatus, producing special medicines, and observing experiments with plants and animals.

Space stations are constructed in space rather than on Earth. This is because the various sections of a space station are too large and too heavy to launch from Earth. It's much easier to take several modules of a station up into space (with a space shuttle, for example) and assemble them in the weightlessness of space than it would be to construct an entire station and launch it. A modular construction allows some parts of the space station to be put together while

Russian space station Mir

leaving room for additional parts as they are developed or when money is available to build them.

One space station proposed by the United States would be assembled over a period of several years using selected modules. Attached to the modules would be several solar panels to gather sunlight and convert it into electricity. One of the modules would be used to house a crew of eight astronauts, whereas other modules would be set aside as laboratories for scientific experiments. Much of the research on this space station would be devoted to the production of very pure drugs and specialized electrical components. The space station might even become a launchpad for journeys to the moon and possibly Mars.

One of the major concerns about life aboard a space station is how the human body reacts when subjected to long periods of weightlessness. In space, muscles and bones are not used in the same way as they are on Earth and they tend to lose some of their functions. In fact, because of weightlessness astronauts are actually taller while they are in space, because the spaces between their vertebrae are not compressed by the force of gravity. What is not known is the effects of this environment on vital organs such as the heart and liver, and the effects on vital functions such as blood circulation. The psychological implications of living in a cramped, weightless environment for long periods also remain unknown.

Space stations offer the potential for increased and extensive exploration of space. Although they are expensive, they do present the possibility of learning more about our world and the other objects in the universe. They can also be used as a "stepping-stone" to further space discoveries.

Rocket Pressure

You'll need:

sheet of construction paper
scissors
tape
straw
clay
cotton swabs
two-liter soda bottle
drill (**Note:** for use by an adult *only*)

What to do:

1. Cut a circle about two inches in diameter from the construction paper. Cut a slit from the edge to the center of the circle. Form the circle into a

cone (cross the edges over each other) and tape the edges together (see the illustration). Make sure there is a small hole in the center (you may need to do some trimming with the scissors).

2. Pull off the cotton from one end of a cotton swab. Push the cotton tip on the other end of the swab up through the hole in the cone so that the cone fits snugly over the cotton tip. This is the "rocket."

3. Ask an adult to drill a hole in the cap of a two-liter soda bottle. The hole should be about the same diameter as the diameter of the straw.

4. Place the straw into the hole in the bottle cap. Seal the straw with clay placed inside the cap.

5. Screw the cap/straw assembly onto the soda bottle tightly.

6. Place the "rocket" into the top of the straw.

7. Point the "rocket" away from people and give the bottle a hard squeeze. The "rocket" will zoom away.

What happens:

When the soda bottle is squeezed, the air molecules inside are compressed, or squeezed together. This increases the air pressure inside the bottle. This air pressure pushes out against the walls of the bottle in every direction. Because the only place this increased pressure can escape is through the opening in the cap, it forces the "rocket" out of the straw with tremendous force. The "rocket" is swiftly propelled into "space."

This project demonstrates a basic law of nature, known as Newton's third law of motion (first postulated by Sir Isaac Newton in 1687). That is, for every action there is an equal and opposite reaction. This means that when you quickly squeeze the soda bottle, you are creating an action, and as a result there must be a corresponding reaction. In this case the reaction is the air rushing out of the bottle and sending the "rocket" out of the straw.

This same principle is used in launching rockets. Rocket fuel provides a force (or an action) that sends a rocket into space (a reaction). In reality, rockets don't use air pressure to go into space; they typically use solid or liquid fuel.

Future Space Exploration Missions

The following chart lists some of the space exploration missions planned for the next several years. Although these missions were on the "drawing board" as this book was being written, please understand that funding, technical modifications, and other factors may postpone some of these missions, modify others, and cancel a few. To get the most current information, log on to the following website and track the status of specific missions: http://nssdc.gsfc.nasa.gov/planetary/chrono_future.html.

Proposed mission	Launch date	Objective
Mars Surveyor	March 7, 2001	Orbit Mars and collect data
Mars Surveyor	April 5, 2001	Land a rover on Mars
SMART 1	Late 2001	Orbit the moon
Muses-C	January 2002	Land on and take samples from an asteroid
CONTOUR	July 2002	Do flyby of three comets
Rosetta	January 23, 2003	Orbit and land on a comet
Champollion/DS4	April 19, 2003	Orbit and land on a comet
Mars Surveyor	May 2003	Acquire samples from Mars
Mars Express	June 2003	Orbit and land on Mars
Europa Orbiter	November 2003	Orbit Europa
Selene	2003	Orbit and land on the moon
Pluto-Kuiper	December 2004	Do a flyby of Pluto
Mars Surveyor	July 2005	Land on Mars and deploy a rover
Mars Surveyor	July 2005	Return samples from Mars
Mercury Orbiter	August 2005	Orbit Mercury

Space Exploration Websites

Following are three fascinating and incredible websites that will provide you with up-to-the-minute news about our exploration of space. You may want to consult these sites on a frequent or regular basis to keep up with the latest happenings and current images of space and our universe.

http://www.spaceviews.com

Spaceviews is a daily digest of the latest information about space exploration. The data here change each day. The site offers news from around the world and across the universe.

http://liftoff.msfc.nasa.gov

Liftoff to Space Exploration has the latest news from the National Aeronautics and Space Administration (NASA), updated each day. Here you can discover what's in the headlines as well as what occurred on each day in space history. This site is jam-packed with space news.

http://photojournal.jpl.nasa.gov

This site provides you with easy access to the publicly released images from various solar system exploration programs. The database currently has more than 1,500 images, with more being added every day. One word describes this site: *COOL!*

What will we discover in space next year . . . in the next decade . . . in the next century? Probably no one knows for sure. But the fact is that there is lots to learn and lots to explore. Maybe you'll be one of those space explorers people will be talking about in two or three centuries. Who knows? It's certainly possible.

9 Learning More About the Universe

The universe is ripe for exploration and rich with an abundance of awaiting discoveries. Even more exciting is the fact that shuttle launches, deep space probes, unmanned explorations of other planets, and a host of scientific findings in space are taking place almost every day.

Many young people are interested in learning more about space. Who knows, perhaps you will be one of the ones to make that first manned exploration of a distant planet. You can begin (or continue) your space explorations by investigating some of the resources listed below.

Space and Universe Groups/Organizations

American Astronomical Society
2000 Florida Ave., NW, Suite 400
Washington, DC 20009
(202) 328-2010

American Institute of Aeronautics and
   Astronautics
The Aerospace Center
370 L'Enfant Promenade, SW
Washington, DC 20024
(202) 646-7444

Association of Astronomy Educators
5103 Burt St.
Omaha, NE 68132
(402) 556-0082

Astronomical Society of the Pacific
390 Ashton Ave.
San Francisco, CA 94112
(415) 337-1100

Challenger Learning Centers
Challenger Center for Space Science
 Education
1029 N. Royal St., Suite 300
Alexandria, VA 22314
(703) 683-9740

National Aeronautics and Space
 Administration (NASA)
Education Division
Code FEE
NASA Headquarters
Washington, DC 20546-0001
(202) 358-0000

NASA Marshall Space Flight Center
Mail Code CL-01
Huntsville, AL 35812-0001
(205) 961-1225

United States Space Foundation
2860 S. Circle Dr., Suite 2301
Colorado Springs, CO 80906-4184
(800) 691-4000

Young Astronaut Council
1308 19th St., NW
Washington, DC 20036
(202) 682-1984

Books

The following books can provide you with loads of information and fascinating facts about the universe and all its dimensions. In these books you'll discover far-away worlds, a cascade of celestial bodies, and an abundance of explorations just waiting to be made. Plan to check out some of these books from your school or public library.

Abernathy, Susan. *Space Machines*. Racine, Wisc.: Western Publishing, 1991.
 Examines the history of space machines from the earliest to the most technical. Excellent overview.

Aronson, Billy. *The Truth behind Shooting Stars: Meteors*. New York: Watts, 1996.
 This book describes the differences between meteors, meteoroids, and meteorites.

Baird, Anne. *The U.S. Space Camp Book of Astronauts*. New York: Morrow, 1996.
 The story of some of America's pioneers in space.

Barrett, Norman S. *The Picture World of Rockets and Satellites*. New York: Watts, 1990.
 Lots of great photographs highlight this colorful introduction to rockets and satellites.

Barrett, Norman S. *The Picture World of Space Shuttles*. New York: Watts, 1990.
 The amazing space shuttle and its history are profiled in this colorful and engaging book.

Barrett, Norman S. *The Picture World of Space Voyages*. New York: Watts, 1990.
This book is a terrific introduction to the space program and what has been discovered as a result of space voyages.

Barton, Byron. *I Want to Be an Astronaut*. New York: HarperCollins, 1989.
A very brief introduction to the life of an astronaut. A good first book.

Beasant, Pam. *1000 Facts About Space*. New York: Kingfisher Books, 1992.
Fun and informative, this book is crammed with exciting facts about the stars, landing on the moon, the history of astronomy, and the future of space discovery.

Becklake, Susan. *All About Space*. New York: Scholastic, 1999.
Loads and loads of information for any future astronaut. This is the book to have!

Becklake, Susan. *Traveling in Space*. Mahwah, N.J.: Troll, 1991.
Filled with lots and lots of data about travel in space, both present and future.

Berliner, Don. *Living in Space*. Minneapolis, Minn.: Lerner, 1993.
How do astronauts go to the bathroom, brush their teeth, or live with others in close quarters? This book explains it all.

Cole, Joanna. *The Magic School Bus Lost in the Solar System*. New York: Scholastic, 1990.
A fantasy adventure in which Miss Frizzle and her class take a wild and crazy ride through the solar system in their magical bus.

Couper, Heather, and Nigel Henbest. *Big Bang: The Story of the Universe*. New York: Dorling, 1997.
Includes theories and explanations about how the universe began.

Couper, Heather, and Nigel Henbest. *Black Holes*. New York: Dorling, 1996.
A voyage of discovery into the mysterious world of black holes.

Fichter, George. *The Space Shuttle*. New York: Watts, 1990.
Contains lots of information about the design and voyages of the space shuttles. Lots of engaging history.

Fraser, Mary Ann. *One Giant Leap*. New York: Holt, 1993.
The voyage and landing of the *Apollo 11* moon mission is told through wonderful illustrations and engaging text.

Gallant, Roy. *The Macmillan Book of Astronomy.* New York: Macmillan, 1986.
A complete guide to all the planets, stars, asteroids, comets, and meteors of our solar system, filled with many photographs and illustrations.

Getz, David. *Life on Mars.* New York: Holt, 1997.
Young readers take a trip to the Red Planet.

Gibbons, Gail. *The Moon Book.* New York: Holiday House, 1997.
Lots and lots of information about the moon; ideal for primary readers.

Gibbons, Gail. *Planets.* New York: Holiday House, 1993.
This is a brief but thorough introduction to the planets, particularly suited for young readers.

Graham, Ian. *The Best Book of Spaceships.* New York: Kingfisher, 1998.
This information-packed book is an ideal reference for any young reader learning about space travel.

Gregory, Valiska. *When Stories Fell Like Shooting Stars.* New York: Simon & Schuster, 1996.
Bear and Fox tell stories about how the moon and the sun fell from the sky.

Gustafson, John. *Planets, Moons and Meteors.* New York: Messner, 1992.
A wonderfully detailed text, filled with photographs and engaging activities, highlights this valuable resource.

Haddon, Mark. *The Sea of Tranquillity.* San Diego: Harcourt Brace, 1996.
A young boy dreams of rocketing to the moon. Finally, on July 19, 1969, he watches history being made.

Harris, Alan. *The Great Voyager Adventure.* Englewood Cliffs, N.J.: Messner, 1990.
The design of the two *Voyager* spacecrafts and discoveries they made are profiled in this intriguing book.

Jones, Thomas, and June English. *Mission Earth—Voyage to the Home Planet.* New York: Scholastic, 1996.
Readers get to ride with an astronaut and a space shuttle crew as they circle Earth for 11 days.

Mason, John. *Spacecraft Technology.* New York: Bookwright Press, 1990.
A good survey of space technology, complemented by loads of excellent photographs.

McConnell, Janet. *Space Travel: Blast-off Day.* Chicago, Ill.: Children's Press, 1990.
> Examines and explains the most basic questions kids have about space travel, in an engaging format.

Milord, Susan. *Tales of the Shimmering Sky.* Charlotte, Vt.: Williamson, 1996.
> A magnificent book that weaves sky tales from around the world with various astronomy activities.

Oxlade, Chris. *Beyond the Night Sky.* New York: Ladybird, 1996.
> See-through pages, wheels, tabs, and flaps offer an interactive view of space.

Petty, Kate. *I Didn't Know That the Sun Is a Star.* Milford, Conn.: Copper Beech, 1997.
> A primary-level book that is full of amazing information about the universe.

Petty, Kate. *I Didn't Know That You Can Jump Higher on the Moon.* Milford, Conn.: Copper Beech, 1997.
> A terrific introduction to space exploration for young readers.

Redfern, Martin. *The Kingfisher Young People's Book of Space.* New York: Kingfisher, 1998.
> Examines our exploration of outer space and discusses the solar system specifically and the universe in general.

Ride, Sally, with Susan Okie. *To Space and Back.* New York: Lothrop, 1986.
> This book answers many questions space enthusiasts may ask. The author, astronaut Sally Ride, describes her own personal adventures in space.

Scagell, Robin. *Space Explained.* New York: Holt, 1996.
> An incredible and delightful reference tool for any youngster looking for the most up-to-date information about the universe.

Scott, Elaine. *Adventure in Space: The Flight to Fix the Hubble.* New York: Disney, 1998.
> Offers young readers a rare look behind the scenes at NASA and the 1993 mission to fix an orbiting telescope.

Scott, Elaine. *Close Encounters: Exploring the Universe with the Hubble Space Telescope.* New York: Disney, 1998.
> Spectacular full-color photographs and computer images highlight this voyage through the history of space exploration.

Simon, Seymour. *Our Solar System.* New York: Morrow, 1992.
A wonderful book that contains up-to-date information about our sun and the planets, moons, asteroids, meteoroids, and comets that travel around it.

Standiford, Natalie. *Astronauts Are Sleeping.* New York: Knopf, 1996.
Beautiful illustrations highlight this story about the thoughts and dreams of three astronauts.

Stott, Carole. *I Wonder Why Stars Twinkle: And Other Questions About Space.* New York: Kingfisher, 1997.
This book presents 30 arbitrary questions about astronomy and space travel. Lots of illustrations.

Trotman, Felicity. *Exploration of Space.* Hauppage, N.Y.: Barrons, 1998.
Cosmonauts and astronauts are the stars of this volume about the history of space exploration.

Vogt, Gregory. *Apollo and the Moon Landing.* Brookfield, Conn.: Millbrook Press, 1991.
Manned explorations to the surface of the moon are detailed in this easy-to-read text.

Vogt, Gregory. *Space Explorers.* New York: Watts, 1990.
Lots of information in an appealing format makes this a most exciting book.

Vogt, Gregory. *Space Stations.* New York: Watts, 1990.
From their earliest beginnings to futuristic designs, space stations are detailed in this delightful book.

Wiese, Jim. *Cosmic Science: Over 40 Gravity-Defying, Earth-Orbiting, Space-Cruising Activities for Kids.* New York: Wiley, 1997.
Fantastic activities and loads of super projects make this book a "must-have" for every young scientist.

Wood, Tim. *Out in Space.* New York: Aladdin Books, 1990.
In this good introduction to the solar system, the reader takes an imaginary trip through space.

Wunsch, Susi. *The Adventures of Sojourner: The Mission to Mars That Thrilled the World.* New York: Mikaya Press, 1998.
An engaging and fascinating tale of the mission to place a remote-controlled rover on the surface of Mars.

Periodicals

The following magazines can provide you with incredible and wonderful information about the universe and space exploration on a regular basis. Check with your parents or guardians about the possibility of obtaining a subscription to one or more of these magazines. You may want to check these out at your local public or school library first to see which one you might enjoy most.

> *Astronomy* (AstroMedia Corporation, 625 E. St. Paul Ave., P.O. Box 92788, Milwaukee, WI 53202)
> This magazine is designed primarily for older students and others who are just beginning to explore the universe.

> *Odyssey* (AstroMedia Corporation, 625 E. St. Paul Ave., P.O. Box 92788, Milwaukee, WI 53202)
> This is an excellent magazine about outer space and astronomy, particularly suited for elementary students.

> *Sky & Telescope* (Sky Publishing Corp., 49 Bay State Rd., Cambridge, MA 02238)
> This magazine is often considered the premier magazine for amateur astronomers.

Websites

The following websites can provide you with valuable background information, a wealth of resources, and numerous tools for studying any area of space science. They can become important in expanding your investigation of the universe as well as in keeping up-to-date with current discoveries and explorations.

Note: These websites were current and accurate as of the writing of this book. Please be aware that some may change, others may be discontinued, and new ones will be added to the various search engines that you use at home or at school.

General

http://www.askanexpert.com/p/ask.html
This site will provide you with opportunities to ask questions of specific scientists, including astronomers and meteorologists.

http://www.nss.org/cyberspace
cyberSPACE News is an online news index produced by the National Space Society. It features links to the latest space-related articles from around the world.

http://www.earthsky.com

Updated daily, this site provides you with the most current data about earth science and astronomy. Be sure to check out the "Calling All Kids" page.

http://explorezone.com/main/space.htm

This site has tons and tons of information about space and space exploration. It's a real winner!

http://spaceplace.jpl.nasa.gov/spacepl.htm

At the *Space Place* you'll get to make spacey things, do spacey things, see space science in action, and learn some amazing facts.

Space Exploration

http://www.jpl.nasa.gov/magellan

Get incredible information and photos about the Magellan mission to Venus (May 1989 to October 1994).

http://www.jpl.nasa.gov/mars

Get up-to-the-minute information about all of NASA's missions to Mars, past, present, and future.

http://quest.arc.nasa.gov/mars

Join the Mars Team Online and get the latest information about the Red Planet, with loads of data and tons of interactive projects.

http://station.nasa.gov/index-m.html

Everything you want to know about the international space station (and more!) can be found at this site. A super site!

http://liftoff.msfc.nasa.gov/academy/academy.html

What does it take to be an astronaut? Join the Space Academy and find out. Everything you'll need to know can be found at this website.

http://ispec.scibernet.com/station

The *Internet Space Station* is one of the most complete Internet resources for young astronauts. Here you'll learn about stars, comets, rockets, space stations, and planets and find fascinating puzzles and challenges. One word sums up this site: *WOW!*

http://www.faahomepage.org

This is one of the coolest sites on the Internet, in which the Future Astronauts of America Foundation provides you with tons of information about every aspect of the universe, from our own planet to the most distant galaxy.

Space History

http://www.nss.org/space/html/history.html

Contains loads of information about the history of space exploration, from its earliest years to the present day.

http://nt.lhric.org/2025/space/menu1.htm

Lots of information about space and space exploration can be found at this site, including a space timeline, project models, and behind-the-scenes information.

http://library.advanced.org/11348/

Galactic Odyssey offers in-depth information about early pioneers and discoveries in the space program. Developed by students.

Space Information

http://www.nss.org/askastro/

The *Ask an Astronaut* website provides the opportunity for you to meet, and then have your space-related questions answered by, those who have been there. Each month a different astronaut is featured.

http://image.gsfc.nasa.gov/poetry/

At this site you can ask specific questions of a working astronomer (Dr. Sten Odenwald). You can also find the answers to more than 3,000 other questions asked by people just like you.

http://www.skypub.com/index.shmtl

The latest information and news releases from *Sky & Telescope* magazine will keep you up-to-date on everything happening in the universe.

http://www.hq.nasa.gov/osf

Here's where you can get the latest information and most up-to-date news about what's happening at NASA and in the space program. This is a great source of news reports and current facts about space exploration.

http://www.spacezone.com

Get up-to-the-minute information and live-action videos about fast-breaking space news. This is real live excitement about space exploration.

http://starchild.gsfc.nasa.gov/docs/starchild/starchild.html

StarChild is a learning center for young astronomers. It's packed with the latest information and lots of fun-to-do stuff.

http://www.seti.org/game

Who's out there? Are humans alone in the universe? Find out the answers to these and other fascinating questions as you roam through this wonderful site, sponsored by SETI (Search for Extraterrestrial Intelligence).

http://www.dustbunny.com/afk

Astronomy for Kids is one of the most user-friendly sites you'll discover on the Web. Full of incredible facts and lots of fun, this is a site you'll come back to time and time again.

Space Science

http://seds.lpl.arizona.edu/nineplanets/nineplanets/nineplanets.html

This website is loaded with an incredible array of space science information, including overviews of all the planets, sound files and video links, and loads of pictures and facts.

http://www.eaglequest.com/~bondono/iconst.html

Expanding Universe is a website that contains a collection of links on astronomy for all grade levels, arranged according to a modified form of the Dewey decimal classification system.

http://www.spaceshots.com

You can get an astronaut's view of the world with poster-sized satellite images taken in space.

http://www.nwlink.com/~sclick/space

This site offers lots of astronomy and space exploration content and access to software reviews, news, and links.

http://www.tcsn.net/afiner

The Nine Planets—For Kids is filled with tons of information about the nine planets of our solar system. It's very "kid-friendly" and is a great place to start your journey through the universe.

Space Shuttles

http://www.icu2.net/faahomepage.gallery1.htm

Everything you want to know about space shuttles, from their earliest years, to how they fly, to what they do on their missions, can be found at this site.

http://seds.lpl.arizona.edu/ssa/docs/spaceshuttle/index.shtml

Filled with loads of history, interesting facts, and super information about the space shuttle, this is a magnificent website that you'll come back to again and again.

http://www.ksc.nasa.gov/shuttle/missions/missions.html

Get all the information you need about every single space shuttle mission. A great resource for school reports.

http://www.cnn.com/tech/space/shuttle.html

How would you like to take off in a space shuttle? Well, here's the site for you! Become an astronaut in real-time simulations that will have you soaring into the skies.

http://www.ksc.nasa.gov

Get live-action videos and real-time shots of the latest events at the Kennedy Space Center. Keep up-to-date on any fast-breaking news or current missions into outer space. It's all here!

Posters and Charts

You may wish to obtain some astronomical posters and charts of the solar system and other celestial objects to post in your room or to share with your classmates. Most are inexpensive and provide wonderful views of various parts of the universe. The following companies offer posters. Write to them and request a copy of their latest catalog.

Celestial Arts
231 Adrian Rd.
Millbrae, CA 94030

Edmund Scientific
101 E. Glouster Pike
Barrington, NY 08007

Nature Company
P.O. Box 7137
Berkeley, CA 94707

Astronomical Society of the Pacific
390 Ashton Ave.
San Francisco, CA 94122

Sky Publishing Company
49 Bay State Rd.
Cambridge, MA 02238

10

More Fantastic Facts

About the Planets

- On Venus, it rains sulfuric acid all the time.
- Every 93,408 years the planets of the solar system assume the same configuration.
- Jupiter radiates more heat than it gets from the sun.
- Jupiter isn't solid and has no solid crust at all.
- More than 1,000 planets the size of Earth could fit into Jupiter.
- Io (one of Jupiter's moons) has a ground surface that billows: It heaves up and down 328 feet every 36 hours.
- One of Jupiter's moons, Sinope, is 14,730,000 miles from the planet.
- Clouds on Venus circle around the planet in four days.
- Earth is two and one-half times larger than Mercury.
- Venus and Mercury have no moons.
- Mercury is the most upright of the planets: Its axis is tilted by just 2 degrees.

Jupiter

Saturn

- Although the seasons on Mars and Earth are similar, martian seasons are twice as long.
- The Great Red Spot, an enormous storm on the surface of Jupiter, is three times larger than Earth.

- Saturn has a storm on its surface (smaller than Jupiter's Great Red Spot) known as Anne's Spot, named after the scientist who discovered it, Anne Bunker.

- So little sunlight reaches Uranus that its average summer and winter temperatures are within 3 degrees F of each other.

- Scientists were not sure if Neptune had rings until the *Voyager 2* spacecraft discovered them in 1989.

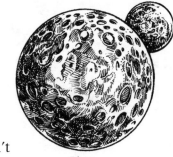

Pluto

- Although Pluto was discovered in 1930, it wasn't until 1978 that its moon, Charon, was discovered.

- The surface temperature on Venus is so hot (860 degrees F) that it can melt lead.

About Space Exploration

- Astronauts on the Apollo moon missions brought back about 882 pounds of moon rocks to study.

- To escape Earth's gravitational pull, a spacecraft must travel faster than 7 miles per second. To escape the sun's gravitational pull would require a speed of 380 miles per second.

- The largest telescope in the world is the Keck Telescope in Hawaii.

- When it returns to Earth's atmosphere, the space shuttle reaches a temperature of 2,300 degrees F.

- The radio telescope at Arecibo, Puerto Rico, can pick up signals from up to 15 billion light-years away.

- Only 12 astronauts have walked on the moon.

- SETI (the Search for Extraterrestrial Intelligence) is an ongoing investigation for signs of life in other places in the universe.

- Spacecraft have flown by all the planets and their moons, except Pluto.

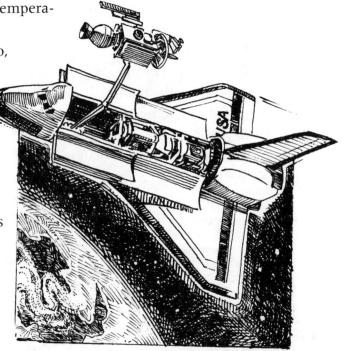

About Stars

- The energy output per second at the sun's core equals the energy released by 90 billion one-megaton hydrogen bombs.
- The closest star to Earth, not counting the sun, is Alpha Centauri, which is 25 trillion miles away.
- The sun reduces in mass by about 4 million tons every second.
- The earliest recorded solar eclipse occurred on October 22, 2137 B.C.
- Every second, on each square yard of its surface, the sun produces enough energy to light 100,000 homes.
- The longest total eclipse of the sun was recorded on June 20, 1955. It lasted for seven minutes, eight seconds.
- The smallest known star measured about 1,000 miles in diameter.
- The sun is about 92 percent hydrogen.
- The center of the sun rotates faster than its poles.
- About 1,300,000 Earths could fit inside the sun.
- A black hole's gravity is so strong that no light can escape from it.
- It is estimated that our sun has been in existence for about 5 billion years and that it will exist for another 5 billion years.
- If the sun were the size of the dot over the letter *i*, the nearest star would be a dot 10 miles away.
- The gases in a sunspot average 3,000 degrees F cooler than the rest of the sun.
- Some stars are 600,000 times as bright as our sun.

About the Moon

- The sun is 100,000 times brighter than the full moon.
- February 1866 was the only month on record without a full moon.
- Every year, it takes the moon two-thousandths of a second more to circle Earth than it took in the previous year.
- The maximum possible duration of a lunar eclipse is 104 minutes.
- Earth is the only planet with a single moon.
- There are about 1,500 moonquakes every year.
- The deepest moon crater is over five miles deep.
- The footprints the astronauts left on the moon will stay there for thousands of years because there is no wind on the moon to blow them away.
- The moon has one-sixth the gravity of Earth. That means that if you weighed 72 pounds on Earth, you would weigh only 12 pounds on the moon.

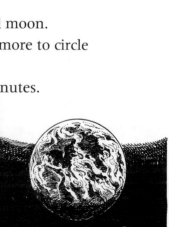

About Earth

- In addition to our satellites, there is about 5 million pounds of space junk circling Earth.
- Every day, meteorites weighing a total of 15,000 tons rain down onto Earth. Most are smaller than a grain of rice.
- There are approximately 455 active volcanoes in the world.
- If Earth were completely smooth, water would cover the entire globe to a depth of about 7,500 feet.
- The pressure at the center of Earth is 27,000 tons per square inch.
- The average thickness of Earth's crust is about 12.5 miles.
- Approximately 75 percent of the rocks on Earth's surface are sedimentary rocks.
- There are only about 120 recognized meteorite craters on Earth, and the oldest is 2 billion years old.
- The continents of Europe and North America are drifting apart by about six feet every 70 years.

About the Universe

- The tail of a comet can extend for more than 90 million miles, the distance between Earth and the sun.
- The known universe weighs about 10 trillion trillion trillion trillion tons.
- The largest known asteroid is about the same size as the state of Texas.
- The Great Comet of 1843 had the longest recorded tail: 205 million miles.
- Our galaxy is about 100 million times wider than our solar system.

Glossary

Asteroid: A small, rocky body, typically found in the asteroid belt between Mars and Jupiter.

Astronaut: A person who travels in space.

Astronomer: A person who studies astronomy.

Astronomy: The study of our universe, galaxy, and solar system.

Atmosphere: The layer of gas that envelops a planet.

Aurora: A display of wavy bands of colored light in the sky that makes Earth's upper atmosphere glow.

Axis: An imaginary line through the center of a planet around which it rotates.

Barred spiral galaxy: A galaxy that has a bar of stars across its center. The arms of the galaxy swing out from the ends of the bar.

Big Bang theory: The premise that the universe was formed as the result of a massive explosion of matter.

Binary star system: A pair of stars that often revolve around each other and are bound by their mutual gravity.

Black hole: The last stage in the life of a massive star in which it collapses in on itself.

Coma: An envelope of gas that surrounds the nucleus of a comet.

Comet: A body of ice and dust that orbits the sun.

Comsat: An artificial communication satellite.

Condensation: The process by which a gas or vapor changes into a liquid.

Constellation: A group of stars that seemingly make a pattern in the sky. There are about 88 recognized constellations.

Crater: A bowl-shaped impression in the surface of a planet, usually caused by the impact of an object from space.

Crust: The outermost solid layer of a planet or moon.

Desalination: The process of removing salts or other chemicals from water.

Dust trail: A stream of dust particles that forms one of two tails of a comet.

Dwarf star: A star of average or less-than-average diameter.

Eclipse: The partial or complete obscuring of one celestial body by another.

Elliptical galaxy: An oval-shaped galaxy.

Erosion: The process by which material is worn away from the surface of an object, such as Earth.

Evaporation: The process by which a liquid changes into a vapor.

Exosphere: The outermost region of a planet's atmosphere.

Galaxy: An enormous group of stars and clouds of gas and dust, held together by gravity.

Gas tail: One of two tails of a comet. This one is straight, narrow, and fainter than the dust trail.

Gibbous phase: The part of the lunar cycle when more than half the moon is visible from Earth.

Gravity: The force that attracts one body to another.

Inner core: The central portion of a planet.

Ionosphere: A region of Earth's atmosphere extending from 31 miles to 600 miles above the surface.

Iris: The Greek goddess of the rainbow and messenger of the gods.

Irregular galaxy: A galaxy with no particular structure or shape.

Landsat: An artificial satellite used to measure and plot land forms on Earth.

Light-year: The distance a beam of light travels in one year.

Local Group: A collection of about 25 galaxies, including the Milky Way.

Magnitude: The degree of brightness of a celestial body, such as a star.

Mantle: The layer of Earth between the crust and the core.

Mass: The measure of the quantity of matter that a body or object contains. It is independent of gravity and thus not related to weight.

Metamorphic: A type of rock that has been altered by heat and pressure.

Meteor: The flash of light produced as a meteoroid passes through Earth's atmosphere.

Meteorite: An object that has survived passage through Earth's atmosphere and has struck the ground.

Meteoroid: A very small body that travels through space.

Meteor shower: A display of a large number of meteors within a short period of time.

Moon: A natural satellite that revolves around a planet.

Navistar: A type of artificial satellite used to plot positions on Earth's surface.

Nebula: A cloudlike patch in the sky made up of gas and dust.

Nuclear fusion: The binding together of atomic particles to build up heavier elements. This causes a release of energy.

Nucleus: The positively charged central region of an atom.

Orbit: The invisible, curved path followed by one object around another object. In space, the orbit of an object is determined by its speed and the gravitational pull between it and the object around which it is orbiting.

Phases: The cycle of changes, as seen from Earth, in the shape of the moon.

Photosphere: The visible outer layer of a star.

Planet: A large body that orbits a star.

Planetarium: A building in which images of celestial bodies can be projected.

Plate tectonics: The theory that semirigid sections of Earth's surface (plates) are in movement. Seismic activity and volcanoes typically take place along the margins of these plates.

Pole: One end of an axis. Planets have two poles.

Prominences: A tonguelike cloud of flaming gases rising from the sun's surface.

Protogalaxy: An initial large cloud of gases, preceeding the formation of a true galaxy.

Protosphere: The sun's visible surface, from which most of the visible light radiates into space.

Protostar: The center of a huge cloud of atmospheric gases. It gets hotter as it shrinks, eventually creating a new star.

Pulsar: The tiny remains of a star that rotate very quickly and send out radio waves.

Red giant: A large star with a relatively cool surface.

Revolution: The motion of a planetary object along its orbit.

Rotation: The spinning of a planet on its own axis.

Satellite: A small body that orbits a larger body in space.

Seasat: An artificial satellite used to track ships at sea.

Sedimentary: Relating to rocks formed by the deposit of sediment.

Shooting star: A meteoroid that passes through Earth's atmosphere and burns up.

Solar distillation: The evaporation and collection of a liquid by condensation. The sun provides the necessary heat to evaporate the water.

Solar flare: A sudden, violent explosion of energy on the sun's surface.

Solar system: The collection of planets and other celestial bodies in orbit around the sun.

Solar wind: The invisible flow of atomic particles from the sun out into space.

Space shuttle: A reusable spacecraft designed to transport astronauts from Earth into space and back.

Space station: A large satellite equipped to support a human crew and designed to remain in orbit for an extended period of time.

Spiral galaxy: A large, flat galaxy with arms in the shape of a pinwheel.

Star: An immense ball of gas that produces large amounts of heat and energy.

Sunspot: A huge magnetic storm appearing as a dark spot on the surface of the sun.

Supernova: A stage in the life of a massive star when the star collapses in on itself and then explodes.

Troposphere: The lowest region of Earth's atmosphere.

Universe: Everything that exists.

Variable star: A star that varies in brightness.

Weathering: A natural process whereby rocks exposed to the weather change in character and break down.

Weightlessness: A state in which the effects of gravity are not experienced.

White dwarf: A small, hot star near the end of its life.

Index